From Here to There and Back Again

Also by Sue Hubbell

A Country Year
A Book of Bees
On This Hilltop
Broadsides from the Other Orders
Far-Flung Hubbell
Waiting for Aphrodite
Shrinking the Cat

from
Here to There
and
Back Again

Sue Hubbell

THE UNIVERSITY OF MICHIGAN PRESS
ANN ARBOR

Copyright © by Sue Hubbell 2004
All rights reserved
Published in the United States of America by
The University of Michigan Press
Manufactured in the United States of America
∞ Printed on acid-free paper
2007   2006   2005   2004   4   3   2   1

No part of this publication may be reproduced,
stored in a retrieval system, or transmitted in
any form or by any means, electronic,
mechanical, or otherwise, without the written
permission of the publisher.

A CIP catalog record for this book is available
from the British Library.

Library of Congress Cataloging-in-Publication Data

Hubbell, Sue.
   From here to there and back again / Sue Hubbell.
     p.   cm.
   ISBN 0-472-11419-0 (alk. paper)
   I. Title

AC8.H89   2004
081—dc22   2004047988

For LeRoy Gilbert,

In Memory

Contents

FOREWORD BY John Michael McGuire *ix*
Preface *xiii*
Annie Sparks *1*
Linnaeus *5*
Delivery Truck *10*
A City of the Mind *16*
Bowling Shoes *20*
The Great American Pie Expedition *24*
The Vicksburg Ghost *52*
Magic in Michigan *71*
Polly Pry *87*
Earthquake Fever *99*
The Honey War *127*
Space Aliens Take Over the U.S. Senate!!! *140*
Blue Morpho Butterflies *153*
The Gift of Letting Go *157*
Mustard *161*
Ozark Springtime *168*

Foreword
John Michael McGuire, *St. Louis Post-Dispatch*

Over the years, if you happened to see a red Chevy pickup moving down a road or highway just about anywhere in this country, chances are that the driver was this wonderful, wandering writer. In the passenger's seat would be her dog, Tazzie (short for Tasmania), a gift from her brother Bil Gilbert, himself a very fine writer. If you honked your horn, Sue Hubbell would have smiled.

Hubbell has roamed about, a real literary nomad and a captivating character. She has lived all over the country, in Michigan, California, Texas, New Jersey, Rhode Island, and now, Maine. The book's title, *From Here to There and Back Again*, captures the nature of Sue Hubbell, once described as a gentle writer of natural history. But as this collection of essays, her eighth book, demonstrates, she's moved well beyond that definition. It reveals her intense interest in a wide range of fascinating things, including Elvis sightings, the quest for the great American pie, earthquakes, supermarket tabloids, bowling shoes, and the world of truckstops. The book also includes personal reflections on the pleasures and pitfalls of running a 90-acre farm in the Missouri Ozarks, and the tangled memories of an "out to lunch" period in her life following the breakup of her first marriage.

That's a glance at the things Hubbell's fingers peck away at. And that pecking has landed her work in an impressive array of publications, including the *New Yorker*, *Time* magazine, the *New York Times*, *Harper's*, *Smithsonian*, and *Sports Illustrated*. To the "Hers" column at the *Times* she contributed Hubbellian work with a peppery feminist tone. She also writes for the *Times'* slick publication *The Sophisticated Traveler*.

Of course, this doesn't touch on her longtime obsession with bugs and flora and fauna, nor on being a beekeeper and making honey, which is why she was known as a gentle writer of natural history. But she is a lot more than this, and among the highlights of my many years as a newspaperman has been watching Sue Hubbell work, and writing about it.

In the late 1980s she was dispatched by *New Yorker* editor Robert A. Gottlieb to Vicksburg, Michigan, where her widowed mother lived. It had been nearly twelve years since Elvis Presley's death, and reports of alleged Elvis sightings in Michigan and elsewhere had made the national news and tabloids, and even drew a TV crew from Australia.

Who better than serious Sue Hubbell (serious in that she'd heard of Elvis, but that was about all) to be shipped to her heartland to write this *New Yorker* piece? She sat in a booth at Mar-Jo's Café in Vicksburg, looking through a misshapen manila pouch, frayed at the edges, notes scribbled on its surface, and containing some Elvis files of the rock and roll historian Joe Edwards, owner of a saloon called Blueberry Hill with a room named after Elvis in the St. Louis suburb of University City. He was a fellow Missourian and good friend of rock legend Chuck Berry. "I think she was stymied until she talked to me," he said.

Given her intense nature, and the fact that she was a librarian at Brown University, Sue Hubbell doesn't stay stymied for very long. She has amazing research abilities, but the author of such books as *A Book of Bees* and *A Country Year: Living the Questions* was someone whose pop-culture interests barely existed. I told her that her doing an article about Elvis was something on the order of Danielle Steele writing a piece on solar telemetry for *Scientific American*. She thought that was funny, and we made the Sunday front-page of the *Kalamazoo Gazette*, our hometown newspaper. That's the way she is. And the range of things that fascinate her seems never-ending.

Take the time a CBS television crew, on assignment for

Charles Kuralt's "Sunday Morning," navigated several doglegs and a tough cutback turn along a network of gravel roads and finally reached her remote bee farm near Mountain View, Missouri. Back then, it was hard to tell if Hubbell watched that much TV. While they were taping this network segment, she began asking the producer about how this or that worked and why. The TV people told her that she should stop in at the CBS studios when she was in New York. And so she did, the next time she and Tazzie drove in the red truck to Bloomingdale's to sell some of her honey.

So who is Sue Hubbell?

Well, after graduating from State High in Kalamazoo, she went off to Swarthmore, and later transferred to the University of Michigan. Typical of the Hubbell way, she continued moving west, earning an undergraduate degree in journalism from the University of Southern California. After that, she motored east again for a master's in library science from Drexel University in Philadelphia. Hubbell now lives in Downeast Maine.

For many years she didn't think she was a writer. Her brother Bil was, she thought, but not her. How wrong that was. She insisted that her first book, *A Country Year*, was merely an exercise in self-examination, a bit of therapy, and not intended for publication. When Random House published it, a reviewer said it was "as elegant and quiet and well-made as one of the small wild things on her Ozark farm that delight this reflective author."

How true.

Who's Sue Hubbell?

"Just so many electrons," she says.

Preface

My family has long had the habit of writing. My brother Bil Gilbert in the next book over, knew from the time he was a child that he was to be a writer and has had a long and distinguished career doing just that. Our cousin John McGuire, also Kalamazoo-raised, is a feature writer at the *St. Louis Post-Dispatch.*

The rest of us have not been so steady. My father, B. LeRoy Gilbert, who was a landscape architect, worked most of his life for the city of Kalamazoo, but a few years before his early death, he began writing a column on landscape gardening for the *Kalamazoo Gazette.* The editor, Jack Walsh, knew a good thing when he saw it and arranged to have the column syndicated. At the time of Roy's death in 1957, the column was running in more than one hundred newspapers.

Then there is shadowy Great Aunt Harriet, a woman with many last names, standing, as I was told when I was a tender youngster, for an equal number of "husbands." She was born in 1838 and in 1908 wrote a short sketch of her life which opened, intriguingly, "Of my own life I do not care to speak in detail. It has been made up of mistakes, ruins, and reverses." Nevertheless, the writing habit was upon her, for even after that demurrer, she went on to add, "When quite young I began writing for different journals east and west. Those stories, sketches, and poems, reviewed in later years seem crude of course."

Would that we were all her equal in honesty and modesty.

Like Bil I began my habit of writing at the *Kalamazoo Gazette* during summer vacations, first from high school and later from college. Bil, a boy person, was assigned to the Sports Desk, but I, a girl person, was put on the Society Desk. The "Society

Pages," a name to make us blush today, were in the long-gone days of the 1950s the equivalent of what we would now call the Style Section, I suppose. After one of those summers, my mother, Marjorie Gilbert, took over my seat at the same desk.

When I was at the University of Michigan, majoring in journalism, I worked several jobs to keep myself afloat financially. One of them was as campus stringer for the *Pontiac Press*. (The other was as a short-order cook at Witham Drugs near the UM campus, and I must confess that I made more money cooking than stringing.) After college I worked at a number of things, none of which had any relationship whatever to writing. One of them was farming in the Ozarks of southern Missouri. In the 1970s I began writing a newspaper column for the *St. Louis Post-Dispatch* in order to eke out the farm income. After that it was one writing thing after another. Some of the results are in the pages that follow.

Annie Sparks

My maternal grandmother, Annie Sparks, lived with our family while I was growing up. When I came home from school, after having made a detour to the kitchen to pour a glass of milk and make a peanut-butter sandwich, I would go up to her sitting room. She would put aside her crossword puzzle—which she did in ink—and tell me family stories while I ate my sandwich.

Her family was like no one else's. My schoolfriends had fathers and grandfathers who *did* things, but in her family the women had been the doers. It wasn't that there weren't menfolk in my grandmother's stories. There were lots of them, but they died young, or were drifters and dreamers who disappeared or turned to drink or succumbed to melancholia or slow mortal diseases. The women, on the other hand, lived a long time and were full of spit and vinegar until the end.

What I learned from her stories was this: women could do hard things and do them competently; problems could be worked out if you ignored what everyone else told you and did what the situation required; sometimes there are men around and sometimes not, but life goes on pretty much the same either way. Those were not bad lessons to learn growing up in the 1930s, when most of the world's messages said something else to a woman-child.

Grandma's own true sweetheart died when she was eighteen, in the eighties of the nineteenth century. After he was gone, she decided to go to college. Her mother was a widow hard-pressed to feed and clothe her eleven children (six lived to adulthood), so Grandma borrowed money from a Man (the word always seemed capitalized when she said it) banker "at 3 percent interest" and went off to a distant college to earn a teaching degree. She held the early woman's movement in contempt because she believed mere equality was a step down for any woman.

She taught school all of her working life. She was a keen bicycler, a good enough tennis player to win a state championship, and a softball player who raged, in her later years, over the ineptness of the Chicago Cubs ("just like a bunch of Men"). She married at thirty-five, was widowed ten years later, and starchily and with considerable airs, raised the two daughters from that brief marriage on a schoolteacher's meager wages. She died, prickly, wary, and protesting, in her nineties.

Her mother's name was Alzina Treat; the name Alzina was supposed to have been mine and I wish it had been, but my mother lost her nerve at the last moment and gave me a more neutral one. I've felt the shadow of that name all my life. I have photographs of Alzina. She had a firm jaw and a fierce eye and in her nineties was still straight and tall, owing to a lifetime habit of preferring the floor to a bed for sleeping. Beds, she claimed, made you soft and self-indulgent—this in an age during which clinics were being set up to get women out of bed, women who had given up, pulled shut the bedroom curtains, and slipped into neurasthenia: Elizabeth Barrett Browning, Olivia Clemens, and Alice James were just the famous ones.

Alzina enjoyed a pipe after supper and her language was too colorful to be quoted in a family newspaper. (Katharine P. Loring, Alice James's companion, once said to her friend, "What an awful pity it is that you can't say 'damn.'")

The first woman in our family about whom I know anything was described at her second trial in London as "willful and corrupt." She was Elizabeth Canning and has been the subject of several books; her criminal career was a hot topic in Europe for a while, engaging the likes of Voltaire and Henry Fielding. She was a servant girl who, on January 1, 1753, disappeared for nearly a month. When found, emaciated, dehydrated, and dazed, she claimed she had been taken captive by gypsies. A gypsy woman was found, tried, and condemned to hang.

But the Lord Mayor of London was suspicious and forced a second trial in which the gypsy woman was exonerated and Elizabeth, the willful and corrupt, was convicted of perjury and transported to the American colonies. There she married a man named Treat, described as "an opulent Quaker" and "a scatterbrained fellow."

My grandmother's best stories were about her aunt, Alzina's sister Harriet. She had been a great beauty and had flaming red hair. She was called, but not to her face, Mrs. H.B.R.B.C.F.E., with an initial for each "husband," some of whom may even have been legal. She lived with one of them, an artist named Bragdon, in New York until one morning, in his studio, he was found "mysteriously dead." When this news reached her family back in Michigan, a thirteen-year-old nephew was sent by train to fetch her back to the Midwest because it was not seemly for a woman to travel alone. That must have amused Harriet, who never gave a fig for seemliness.

But she did return, began accumulating husbands, but supported herself with a wholesale stationery business in Chicago. When her young tubercular brother (about whom Grandma Annie always said proudly, "And he had a poem published in the *Atlantic Monthly*") was taken prisoner in the Civil War and lay languishing in a Southern jail, Harriet locked up her stationery warehouse and betook herself to Washington. There she talked President Lincoln into a corner and made him agree to a prisoner exchange so that this pale, delicate young man could come home to die. My mother and her sister claimed to have a letter from Lincoln laying all this out, but they never did produce it.

Are these stories true? Some of them go back so far that, at very least, they must have improved in the telling. Like Elizabeth Canning, all of us enjoy telling stories. But their accuracy

Annie Sparks

doesn't make any difference. These are the myths that tell us what it is like to be a woman in my family. These were determined, self-assured, outrageous women. They were fierce in their need to take charge and fierce in their enthusiasms. They were "difficult" women. My English ancestors had a strong blend of Scotch-Irish blood in their veins, and there is an old Scottish prayer that fits them well: Oh Lord, grant that I may always be right for Thou knowest how hard I am to turn.

Their stories must have been on my mother's mind when, widowed in her fifties, pensionless, and without job skills, she first began working for a living. They were probably also there when she was forced to retire at sixty-five. "Hmpf!" she said. "I'll show them who's fit." She joined the Peace Corps and went off to India to spend what were probably the best years of her life.

The stories have always haunted me. When I was young those women embarrassed me, and I would try to be sweet and agreeable as long as I could stand it. I labored to escape the pattern. I married a university professor, raised a son, and worked as an academic librarian. My husband and I moved to the Ozarks, bought a farm, and started a commercial beekeeping business. And divorced. So, in my forties, I found myself alone in possession of a farm, with debts so big and work so hard I would wake up at night in a panic to stare at the ceiling, then remember those women and how they had lived their lives. Growing in crotchets and cussedness with every passing year, I managed and became gloriously happy in the process.

I am fifty-three now and have made peace with the fact that I, too, am a difficult woman, and would rather have my stories than anyone else's.

—*New York Times*, May 27, 1988.

Linnaeus

I met Paul, the boy who was to become my husband, when he was sixteen and I was fifteen. We were married some years later, and the legal arrangement that is called marriage worked well enough while we were children and while we had a child. But we grew older, and the son went off to school, and marriage did not serve as a structure for our lives as well as it once had. Still, he was the man in my life for all those years. There was no other. So when the legal arrangement was ended, I had a difficult time sifting through the emotional debris that was left after the framework of an intimate, thirty-year association had broken.

 I went through all the usual things: I couldn't sleep or eat, talked feverishly to friends, plunged recklessly into a destructive affair with a man who had more problems than I did but who was convenient, made a series of stupid decisions about my honey business, and pretty generally botched up my life for several years running. And for a long, long time, my mind didn't work. I could not listen to the news on the radio with understanding. My attention came unglued when I tried to read anything but the lightest froth. My brain spun in endless, painful loops, and I could neither concentrate nor think with any semblance of order. I had always rather enjoyed having a mind, and I missed mine extravagantly. I was out to lunch for three years.

 I mused about structure, framework, schemata, system, classification, and order. I discovered a classification Jorge Luis Borges devised, claiming that

A certain Chinese encyclopedia divides animals into:

a. *Belonging to the Emperor*
b. *Embalmed*
c. *Tame*
d. *Sucking pigs*
e. *Sirens*
f. *Fabulous*
g. *Stray dogs*
h. *Included in the present classification*
i. *Frenzied*
j. *Innumerable*
k. *Drawn with a very fine camel-hair brush*
l. *Et cetera*
m. *Having just broken the water pitcher*
n. *That from a long way off look like flies.*

Friends and I laughed over the list, and we decided that the fact that we did so tells more about us and our European, Western way of thinking than it does about a supposed Oriental world view. We believe we have a more proper concept of how the natural world should be classified, and when Borges rumples that concept it amuses us. That I could join in the laughter made me realize I must have retained some sense of that order, no matter how disorderly my mind seemed to have become.

My father was a botanist. When I was a child he reserved Saturday afternoons for me, and we spent many of them walking in woods and rough places. He would name the plants we came upon by their Latin binomials and tell me how they grew. The names were too hard for me, but I did understand that plants had names that described their relationships one to another and found this elegant and interesting even when I was six years old.

So after reading the Borges list, I turned to Linnaeus. Whatever faults the man may have had as a scientist, he gave us a beautiful tool for thinking about diversity in the world. The first word in his scheme of Latin binomials tells the genus, grouping diverse plants which nevertheless share a commonality; the second word names the species, plants alike enough to regularly interbreed and produce offspring like themselves. It is a framework for understanding, a way to show how pieces of the world fit together.

I have no Latin, but as I began to botanize, to learn to call the plants around me up here on my hill by their Latin names, I was diverted from my lack of wits by the wit of the system.

Commelina virginica, the common dayflower, is a rangy weed bearing blue flowers with unequal sepals, two of them showy and rounded, the third hardly noticeable. After I identified it as that particular *Commelina*, named for a sample taken in Virginia, I read in one of my handbooks, written before it was considered necessary to be dull to be taken seriously:

> Delightful Linnaeus, who dearly loved his little joke, himself confesses to have named the day-flowers after three brothers Commelyn, Dutch botanists, because two of them—commemorated in the showy blue petals of the blossom—published their works; the third, lacking application and ambition, amounted to nothing, like the third inconspicuous whitish third petal.

There is a tree growing in the woodland with shiny, oval leaves that turn brilliant red early in the fall, sometimes even at summer's end. It has small clusters of white flowers in June that bees like, and later blue fruits that are eaten by bluebirds and robins. It is one of the tupelos, and people in this part of the country call it black-gum or sour-gum. When I was growing up in Michigan I knew it as pepperidge. Its botanic name is *Nyssa sylvatica*. *Nyssa* groups the tupelos, and is derived from the Nyseides—the Greek nymphs of Mount Nysa who cared for the infant Dionysus. *Sylvatica* means "of the woodlands." *Nyssa sylvatica*, a wild, untamed name. The trees, which are often hollow when old, served as beehives for the first American settlers, who cut sections of them, capped them, and dumped in the swarms that they found. To this day some people still call beehives "gums," unknowingly acknowledging the common name of the tree. The hollow logs were also used for making pipes that carried salt water to the salt works in Syracuse in colonial days. The ends of the wooden pipes could be fitted together without using iron bands, which would rust.

This gives me a lot to think about when I come across *Nyssa sylvatica* in the woods.

I botanized obsessively during that difficult time. Every day I learned new plants by their Latin names. I wandered about the woods that winter, good for little else, examining the bark of leafless trees. As wildflowers began to bloom in the spring, I carried my guidebooks with me, and filled a fat notebook as I identified the plants, their habitats, habits, and dates of blooming. I had to write them down, for my brain, unaccustomed to exercise, was now on overload.

One spring afternoon, I was walking back down my lane after getting the mail. I had two fine new flowers to look up when I got back to the cabin. Warblers were migrating, and I had been watching them with binoculars; I had identified one I had never before seen. The sun was slanting through new leaves, and the air was fragrant with wild cherry (*Prunus serotina: Prunus*—plum, *serotina*—late blooming) blossoms, which my bees were working eagerly. I stopped to watch them, standing in the sunbeam. The world appeared to have been running along quite nicely without my even noticing it. Quietly, gratefully, I discovered that a part of me that had been off somewhere nursing grief and pain had returned. I had come back from lunch.

Once back, I set about doing all the things that one does when one returns from lunch. I cleared the desk and tended to the messages that others had left. I had been gone for a long time, so there was quite a pile to clear away before I could settle down to the work of the afternoon of my life, the work of building a new kind of order, a structure on which a fifty-year-old woman can live her life alone, at peace with herself and the world around her.

>—From *A Country Year: Living the Questions* by Sue Hubbell. Copyright © 1983 by Sue Hubbell. Reprinted by permission of Houghton Mifflin Company. All rights reserved.

Delivery Truck

When I come to New York, I stay at the YMCA on Forty-seventh Street between Second and Third. The Y is a liberal and welcoming sort of place. I am not Y, I am not M, I am not C, and yet I am allowed to stay there, and even use the swimming pool. The location is convenient. The place is clean and cheap: twenty-eight dollars a night. Of course, the rooms are small—the size I like to imagine monks' cells to be—but, after all, I'm only in my room to sleep. The bathrooms are shared, but that is rather Continental. And lots of Europeans stay there. In the elevators, I hear spirited conversations in a variety of tongues. I can also leave my delivery truck out in front, because the whole block is designated as a truck-standing zone. I'm not sure how *long* I can leave it there. I once parked, went into the Y, checked in, unpacked, took a shower and washed my hair, and came out to find my truck unticketed. So a truck-loading zone is good for at least that long.

Another thing I like about the Y is that there is a Smiler's all-night deli just around the corner, on Third, so when I get up at five in the morning, I can walk over there, buy a cup of coffee to go, bring it back, sit in my tiny little room, and drink it while I plan my delivery route. I like to be over at Macy's loading dock, on Thirty-fifth Street, before six-thirty. The men who work there have all come in by then and are having their coffee and Danish, and they aren't yet exasperated, as they sometimes are after the big semis start to arrive. My truck is big as pickups go—it's a three-quarter ton—but it's a lot smaller than a tractor-trailer rig, which is what most loading docks are set up for. My deliveries, of just a few cartons of honey at a time, are a nuisance to most dock men. But at six-thirty the men of Macy's are still in a good mood. They let me back up to one of the bays. I put down my tailgate, which is several feet below the loading-dock level, and unload my cartons up onto pallets, which they will move with a forklift. In winter, it's still dark then, and I have to wander around the loading dock in golden lamplight to find someone who will sign my invoice and officially receive my delivery. He checks the purchase order number and the department destination number, stamps the invoice in several places, and scrawls his initials on it. I give him the duplicate copy, and in several weeks, perhaps a month or longer, I'll receive a check for what I've delivered.

If I'm not there by six-thirty, I go looking for the dispatcher, who has a booth on the street and is sometimes in it. The first time I delivered to Macy's, he noticed my Missouri license plates. With a sigh, he told me how much he missed his girlfriend, who lived in Kansas City and was stubborn about moving to New York to be with him. We talked about Missouri and the recalcitrance of lovers for a while. He helped me find a parking place and let me wheel my cartons up a walk-in ramp on my hand truck. He remembers me from year to year and always gives me a big hug.

After Macy's, I drive to Seventh, hang a left, turn left again on Thirty-fourth, right on Broadway, and go speeding downtown to the World Trade Center. It's enormous fun to drive down Broadway at that hour of the morning. No one is about except other delivery drivers, and we all hustle right along, making every light, the big wheels on our trucks scarcely noticing the potholes. New York is dark, cozy, and ours at that time of day.

Deliveries at the World Trade Center are made underground, down a ramp off West Broadway, in a cavernous space with a dock running the length of it. At the entrance, a guard looks over my invoice, checks to see that I have commercial license plates, and waves me on to what he says is an appropriate area. I thread my way between big pillars, tractor trailers, and a few vans. After I'm parked, it gets confusing and uncomfortable. No matter what the outside temperature, it's warm and clammy there, and the air is bad. After I've loaded my cartons on my hand truck, I'm drenched in sweat. The rules change every time I go there. My deliveries end up in Inhilco's Big Kitchen, on the concourse, but sometimes I have to take my cartons up in the freight elevator all the way to an Inhilco storage room, on the 106th floor—the ride is so fast it makes my ears pop—and then trundle my hand truck around until I can find someone to sign for the delivery. If I'm lucky, though, I can leave my cartons on the loading platform and one of the workers will take them up.

The workers are, for the most part, young men who speak little English. I speak little Spanish. Fortunately they bring forth all their Latin courtesy for an older woman, and somehow, with many gestures and in phrases halfway between our languages, we communicate. They are always sweet and helpful, and do what they can to save me from lifting my cartons, which weigh forty pounds each. One day, I had been sent from one uncomprehending young man to another. My sweatshirt was soaked, my hair disheveled. I was tugging my hand truck behind me, clutching my invoice book to my chest. My purse dangled from my shoulder. A young man stopped me, his dark eyes melting with concern, to point out that my purse was unzipped and a twenty-dollar bill was hanging from it. "You must be careful in this big city," he said, haltingly.

Balducci's is next. I've sold to Balducci's for years, and remember the first time, when Louis himself came out to the truck, which was triple-parked on Sixth Avenue, to dicker over the price and help unload the order. The store has grown too prosperous for that now. The stock is controlled by a computer, and I deliver to a storage area on Eleventh Street between First and Second, where there are no spaces reserved for trucks unloading. So I have to double-park and block all the early-morning traffic behind me while I ring a concealed bell and wait for a man to open a heavy metal grillwork gate. He usually comes out to help me unload, however, and once the drivers of the cars trying to get through see what is happening they settle down to honking in only a perfunctory way.

Next, I head uptown, to deliver to Bloomingdale's and a number of small shops on the East Side. The traffic is beginning to pick up by then, but in my truck I am high above it and can plan my strategy, threading my way in and out, going as fast as possible to finish my deliveries before ten-thirty, when double-parking, even with blinkers flashing, becomes unconscionable.

The little shops usually only take one carton, and so those deliveries are quick if I plan my route carefully and avoid as much as possible the narrow cross streets where I have to suck in my mirrors (by folding them back), pray to the trucking gods that a tank truck will not be delivering fuel oil, and squeeze by.

At Bloomingdale's, I leave my truck running on Fifty-ninth Street—its flashers going and tailgate down, so that the traffic wardens in brown can see what I am doing—while I go inside to the gourmet food section to find someone willing to authorize the delivery. After this person has satisfied himself of my good intentions, he opens a lock on a very heavy metal trapdoor on the sidewalk outside the Fifty-ninth Street entrance. I go back and tug it open, sometimes only with the help of a strong passerby. Then I slip my cartons down the chute. I try to close the trapdoor in a satisfactory way, but I often fail, and leave the edge sticking up. This makes the man inside exceedingly cross. He invariably reminds me that people could sue Bloomingdale's if they tripped over the door, and that it would be my fault.

Once, the Bloomingdale's buyer directed me to deliver to the warehouse, over in Long Island City. I changed my schedule around and, after driving down from Boston on a wintry day, planned to stop at the warehouse before checking into the Y.

After losing my way several times, I found the warehouse at about 1:00 P.M. and pulled into a line of small trucks and delivery vans. I waited for half an hour and nothing happened, so I got out and asked what the protocol was. The other drivers, all men—Haitians, Puerto Ricans, Italians, Jamaicans, and a black man from southern Alabama—were helpful and friendly. They advised patience. They explained that I, like them, was a "straight trucker." (This refers to the way we back our trucks up. We do it straight; tractor-trailer rigs cut and bend.) There were a lot of loading bays, but only one for straight truckers. Each load had to be processed on a computer, whose speed was that of geological change, apparently—the lucky straight trucker who was now at the bay had been there nearly an hour and had only three pieces to unload.

A Jamaican driver explained that he had been waiting since eight in the morning and had just a single carton to deliver. He'd tried to walk it in, but the dock workers had barred his way. "All closed down now anyway," he said in his lilting English. "The guys inside—they have stopped for lunch."

So I got back in my truck and waited. I took my checkbook out of my purse and balanced it. I planned my delivery route for the next day. On a tiny notepad I wrote a sixteen-page letter to a friend. Two hours had passed. I had been able to inch forward one truck length. I got out and stretched, talked to the other drivers. They told me that the area was very unsafe—that young hoodlums would break into waiting trucks and rob them if the drivers left them. They advised me to keep my truck locked. We talked some more. Christmas was not far off, and one driver, recently arrived from Puerto Rico, told me about the beautiful doll he was planning to buy for his daughter back on the island. I gave him a jar of honey to send along. Then I gave out jars of honey to the rest of the drivers. It was getting cold, and we all stamped our feet to stay warm. We moved our trucks another length and complained about Bloomingdale's computer, about computers in general.

SUE HUBBELL

We were all getting hungry. Lunch was over for the men inside the warehouse, but none of us had had ours. The Alabamian knew of a Greek restaurant several blocks away, but everyone agreed that it was unsafe for us to leave our trucks, even though they were locked. The men looked around, and decided that I seemed to be reliable. (There is an advantage to growing old and becoming motherly-looking.) Everyone dug into his pockets, we put together a kitty, and I took orders: burgers with various fixings, fries, and coffee in differing degrees of lightness and sweetness. I set off on foot for the Greek restaurant and found it after a wrong turn or two. The counterman cooked our burgers to order and poured our coffees, and, hugging the warm sacks to me, I walked back against the cold wind to the drivers. We spread our feast out on a hood top and ate and talked. We talked trucks—the good points of some, the defects of others. We talked driving and compared hard places for deliveries. We talked families. They showed me pictures of their babies, and I told them about my son in Boston. Occasionally, we got back into our trucks and inched up in line. We talked about what it is like to come to New York from far places and what those far places are like and what they mean to us.

At seven in the evening I was finally at the unloading bay. I slid out my cartons of honey, the computer processed them, a dock worker signed my invoice, I drove out and waved to the remaining truckers. It was one of the best days I've ever had in New York.

—*The New Yorker*, March 7, 1988.

Delivery Truck

A City of the Mind

In the early morning there is a city of the mind that stretches from coast to coast, from border to border. Its cross streets are the interstate highways, and food, comfort, companionship are served up in its buildings, the truck stops near the exits. Its citizens are all-night drivers, the truckers, and the waitresses at the stops.

In daylight the city fades and blurs when the transients appear, tourists who merely want a meal and a tank of gas. They file into the carpeted dining rooms away from the professional drivers' side, sit at the plastic-topped tables set off by imitation cloth flowers in bud vases. They eat and are gone, do not return. They are not a part of the city and obscure it.

It is 5:00 A.M. in a truck stop in West Virginia. Drivers in twos, threes, and fours are eating breakfast and talking routes and schedules.

"Truckers!" growls a manager. "They say they are in a hurry. They complain if the service isn't fast. We fix it so they can have their fuel pumped while they are eating and put telephones on every table so they can check with their dispatchers at the same time. They could be out of here in half an hour. But what do they do? They sit and talk for two hours."

The truckers are lining up for seconds at the breakfast buffet (all you can eat for $3.99—biscuits with chipped-beef gravy, fruit cup, French toast with syrup, bacon, pancakes, sausage, scrambled eggs, doughnuts, Danish, cereal in little boxes).

The travel store at the truck stop has a machine to measure heartbeat in exchange for a quarter. There are racks of jackets, belts, truck supplies, tape cassettes. On the wall are paintings for sale, simulated wood with likenesses of John Wayne or a stag. The rack by the cash register is stuffed with Twinkies and chocolate Suzy Qs.

It is 5:00 A.M. in New Mexico. Above a horseshoe-shaped counter, on panels where a menu is usually displayed, an overhead slide show is in progress. The pictures change slowly, allowing the viewer to take in all the details. A low shot of a Peterbilt, its chrome fitting sparkling in the sunshine, followed by one of a bosomy young woman, the same who must pose for those calendars found in auto parts stores. She almost has on clothes and she is offering to check a trucker's oil. The next slide is a side view of a whole tractor-trailer rig, its eighteen wheels gleaming and spoked. It is followed by one of a blonde bulging out of a hint of cop clothes writing a naughty trucker a ticket.

The waitress looks too tired and too jaded to be offended. The jaws of the truckers move mechanically as they fork up their eggs-over-easy. They stare, glassy-eyed, as intent on chrome as on flesh.

It is 4:00 A.M. in Oklahoma. A recycled Stuckey's with a blue tile roof calls itself simply Truck Stop. The sign also boasts showers, scales, truck wash, and a special on service for $88.50. At a table inside, four truckers have ordered a short stack and three eggs apiece, along with bacon, sausage, and coffee (Trucker's Superbreakfast—$3.79).

They have just started drinking their coffee, and the driver with the Roadway cap calls over the waitress, telling her there is salt in the sugar he put in his coffee. She is pale, thin, young, has dark circles under her eyes. The truckers have been teasing her, and she doesn't trust them. She picks up a canister, dabs a bit of sugar from it onto a finger, tastes it. Salt. She samples sugar from the other canisters. They have salt too, and she gathers them up to refill them. Someone is hazing her, breaking her into her new job. Her eyes shine with tears.

She brings the food and comes back when the truckers are nearly done. She carries a water jug and coffeepot on her tray. The men are ragging her again and her hands tremble. The tray falls with a crash. The jug breaks. Glass, water, and coffee spread across the floor. She sits down in a booth, tears rolling down both cheeks.

"I'm so tired. My old man . . . he left me," she says, the tears coming faster now. "The judge says he's going to take my kid away if I can't take care of her, so I stay up all day and just sleep when she takes a nap and the boss yells at me and . . . and . . . the truckers all talk dirty . . . I'm so tired." She puts her head down on her arms and sobs luxuriantly. The truckers are gone, and I touch her arm and tell her to look at what they've left her. There is a twenty-dollar bill beside each plate. She looks up, nods, wipes her eyes on her apron, pockets the tips, and goes to get a broom and a mop.

It is 3:30 A.M. in Illinois at a glossy truck stop that offers all mechanical services, motel rooms, showers, Laundromat, game room, TV lounge, truckers' bulletin board, and a stack of newspapers published by the Association of Christian Truckers. Piped-in music fills the air.

The waitress in the professional drivers' section is a big motherly-looking woman with red hair piled in careful curls on top of her head. She correctly sizes up the proper meal for the new customer at the counter.

"Don't you know what you want, honey? Try the chicken noodle soup with a hot roll. It will stick to you like you've got something, and you don't have to worry about grease."

She has been waitressing forty years, twenty of them in this truck stop. As she talks she polishes the stainless steel, fills mustard jars, adds the menu inserts for today's special (hot turkey sandwich, mashed potatoes and gravy, pot of coffee—$2.50).

"The big boss, well, he's a love, but some of the others aren't so hot. But it's a job. Gotta work somewhere. I need a day off though. Been working six, seven days straight lately. Got shopping to do. My lawn needs mowing."

Two truckers are sitting at a booth. Their faces are lined and leathery. One cap says HARLEY-DAVIDSON, the other COORS. Harley-Davidson calls out, "If you wasn't so mean, Flossie, you'd have a good man to take care of you and you wouldn't have to mow the damn lawn."

She puts down the mustard jar, walks over to Harley-Davidson and Coors, stands in front of them, hands on wide hips. "Now you listen here, Charlie, I'm good enough woman for any man, but all you guys want are chippies." Coors turns bright red. She glares at him. "You saw my ex in here last Saturday night with a chippie on his arm. He comes in here all the time with two, three chippies just to prove to me what a high old time he's having. If that's a good time, I'd rather baby-sit my grandkids."

Chippies are not a topic of conversation that Charlie and Coors wish to pursue. Coors breaks a doughnut in two, and Charlie uses his fork to make a spillway for the gravy on the double order of mashed potatoes that accompanies his scrambled eggs.

Flossie comes back to the counter and turns to the new customer in mirror shades at this dark hour, a young trucker with cowboy boots and hat. "John-Boy. Where you been? Haven't seen you in weeks. Looks like you need a nice omelet. Cook just made some of those biscuits you like too."

I leave a tip for Flossie and pay my bill. In the men's room, where I am shunted because the ladies' is closed for cleaning, someone has scrawled poignant words: NO TIME TO EAT NOW.

—*Time Magazine*, June 3, 1985.

Bowling Shoes

A friend of mine found these terrific shoes in a secondhand store, bowling shoes, divided longitudinally by color: The outer half is red, the inner half is blue, and a stripe of white runs up from the toe to tongue. They have a pair of crossed bowling pins on each heel and a pair of jingle bells wired onto the laces. The shoes weren't her size, but she bought them anyway, and then gave them to me, with a smile, when she found out they would fit me. I wear them a lot. I like to look down at them twinkling away there on my feet; they make me feel like doing soft-shoe routines. I went out to dinner with friends the other evening and we found that the shoes knew how to gallop, a skill we had learned in second-grade phys ed but had forgotten.

I was in St. Louis not long ago, and the shoes walked me right into the National Bowling Hall of Fame and Museum. It's across the street from Busch Stadium—a pudgy building of gray and pink pebbles firmly anchored on a triangular lot by a squat tower at each apex. Inside are three floors of bowling displays and trinkets, and eight functioning bowling lanes. St. Louisans rent the place after hours for parties. They hire a disc jockey and a caterer and spend the evening eating, dancing, looking at the exhibits, riding the elevators, and when all else palls they bowl a few frames. Last year, more than two hundred events were held at the museum, including a fortieth-birthday party, a wedding reception, and a bar mitzvah.

The display area is along an easy ramp that spirals through the building, and on it I learned, from a diorama showing a slouch-faced humanoid dressed in hides and animal teeth, that the first bowler may have been a caveman tossing rocks. A glass case held reproductions of bowling pins from 5200 B.C., unearthed in Egypt by Sir Flinders Petrie. They looked like pointed darning eggs with flat bottoms. A panel showed two muscly Anglo-Saxons escaped from an opera stage and dressed in skins and horned caps; one was holding a thick cup of something refreshing, and the other, war ax strapped to his waist, was holding an immense skin ball over his head. According to the panel, the Brothers Grimm and the historian Wilhelm Pehle traced bowling back to a pre-Christian game in northern Europe called *Steinstossen*, which consisted of knocking down upright stones by throwing weapons at them. Pehle went on to link bowling to a church ritual among the northern tribes in which parishioners rolled a stone at a *kegel*, or a club used for self-defense, which they renamed *heide*—the heathen, that which is pagan, the Devil. Toppling the *heide* symbolized the slaying of Satan, according to St. Boniface, an eighth-century English missionary, who adapted the game for his flock.

Strong stuff! But my shoes jingled me on up the ramp and planted themselves in front of a diorama of Martin Luther bowling. There he is, life-size, in flowing black robes, a soft hat on his head, and with a sprightly touch of red at his throat. His feet are in the correct bowling stance, and his hand is back and curved over a ball, which he is about to throw at wooden pins. Two children crouch watching. Behind him are bowling banners. Martin Luther built a bowling alley for his family and household, and he wrote that he himself enjoyed rolling out a ball now and then.

Apparently all over Europe there were people having too much fun bowling, for eventually the Germans, the French, and the English all banned the game, ostensibly because it encouraged gambling and rowdyism, separated families, and lured men from the armies. But the laws made exceptions for the wealthy and the titled, and members of the lower orders found loopholes. Sly bowlers in England invented a game, not specifically banned by law, called bubble the justice, in which balls were rolled into shallow holes. The museum's message is that the urge to bowl is irrepressible and that all over Europe people went on bowling and called what they were doing by another name: *skales, loggats, kayles, cloish, Platzbahnkegeln, quilles de neuf, skittles.*

In America, the English brought lawn bowling to Virginia. The Dutch brought ninepins to New York. At the end of the spiral ramp, I stopped to peer into a mutoscope, an elegant gilt-spangled, light-filled stand the size and roughly the shape of a large gumball machine—a turn-of-the-twentieth-century penny-arcade version of the movies. It told the story of Rip Van Winkle on a roll of cards. When I turned the mutoscope's crank fast enough, Rip's diminutive friends spiritedly knocked down their ninepins. At an IBM terminal at the top of the museum, even I, a computer illiterate, could press enough keys to discover that no one in the town where I live has ever bowled a three hundred, or perfect, game, but that in the town in which I was born five screenfuls of folks have. I knew none of them. I felt like Rip Van Winkle.

My shoes danced on through the Bowling Hall of Fame (dark, respectful, male) and the Women's Bowling Hall of Fame (light, cheerful, suggestive of cookies). The museum was funded by major donations from equipment manufacturers and the producers of those beverages usually associated with the sport, and also by donations from bowling associations, many of which raised their contributions through bake sales. I examined a bowling-shirt display, antique bowling pins, bowling in art and quilt and stein. I circled the bowling lanes (four modern and four nostalgic); I would have liked to use one of them, but my shoes reminded me I didn't know how. I admired the historic-beer-tray display and ended up at the world's only mobile bowling pin, built around the chassis of a 1936 Studebaker coupe.

On my way back out to the street, I waved to the young man at the visitors' desk.

"*Nice* shoes," he said with a smile.

—The New Yorker, January 23, 1989.

The Great American Pie Expedition

> Nobody needs drug-store ice cream;
> pie is good enough for anybody.
>
> —*Main Street*, by Sinclair Lewis, 1920

Easy as pie. It's easy to make good, tasty, comforting, serviceable pie. Of course, excellent pie is harder to make, and harder to find, but excellence is always rare. The Shaker Lemon Pie served at Shaker Village of Pleasant Hill, near Harrodsburg, Kentucky, is an excellent pie. The Shakers invented this pie back in the early 1800s, when they began trading goods they grew or manufactured for the few necessities they couldn't produce. Lemons, which they considered an important item in a healthy diet, were one of the "world's goods" they needed. Their lemons came all the way from New Orleans and were so dear that the Shakers believed it a sin to waste any part of them, so they devised a recipe that would use the whole lemon.

Pie from that recipe is served in the public dining room of the Trustees' House at Shaker Village. The pie has a beautifully browned crust—surprisingly thick, but so light and flaky that it shatters under a fork. Inside is a generous lemon custard, with bits of lemon pulp and rind throughout. The sourness of the lemon plays off the sweetness of the custard in an altogether delicious way. It is a pretty pie, too. Here is the way to make both the crust and the filling. The recipes are from *We Make You Kindly Welcome*, a cookbook available at the restaurant.

CRUST FOR A 9-INCH PIE
2 C. flour
1 tsp. salt
⅔ C. plus 4 tbsp. shortening
4 tbsp. cold water

Mix flour and salt. Cut shortening into flour until it forms very small balls. Sprinkle in water, a tbsp. at a time, while mixing lightly with a fork until the flour is moistened. Mix it into a ball that cleans the bowl. Do not overwork the dough. Roll out on floured board.

(The less you handle dough, the better; handling toughens it. A good crust must be crisp, even at the risk of being a bit crumbly.)

SHAKER LEMON PIE
2 large lemons with very thin rinds
4 eggs, well beaten
2 C. sugar
Crust for pie

Slice lemons as thin as paper, rind and all. Combine with sugar; mix well. Let stand 2 hours, or preferably overnight, blending occasionally. Add beaten eggs to lemon mixture; mix well. Turn into 9-inch pie shell, arranging lemon slices evenly. Cover with top crust. Cut several slits near center. Bake at 450° for 15 minutes. Reduce heat to 375° and bake for about 20 minutes or until silver knife inserted near edge of pie comes out clean. Cool before serving.

Pleasant Hill is on U.S. 68, the Harrodsburg Road, which winds past prosperous-looking horse farms, curving through stratified rock as it dips down to cross the Kentucky River. There have been no Shakers at Shaker Village since 1923, but it has been restored to its bucolic picturesqueness. After eating the excellent lemon pie, I went back to my car and let my dog, Tazzie, out for a run. We found a sheepy hillside and sprawled out on it, soaking up the early-spring sunshine. Tazzie, who is mostly German shepherd but with soft edges, gazed decorously at the sheep, sniffed the air, and rolled over on her back in the short grass, kicking her feet lazily in the air. I lay on my back, too, staring at fluffy clouds and listening to meadowlarks.

My family has always been a pie family. When my son, Brian, was in high school, he spent more time hitchhiking around the country with friends than I or those in loco parentis at his boarding school ever knew. He and his friends seldom had much money, but they always had enough in their jeans to buy pie and coffee. Fifteen years and more later, he is still an expert on pie in astonishing places. A few years ago, I was driving across the country with him and his wife, Liddy, when, in a small town in Kentucky, an abstracted look came across his face. He said to Liddy, who was driving, "If you'll turn left at the stop sign and then go two blocks, there's a place that has really good pecan pie." We did, and there was. We sat there eating pie and drinking coffee, and, caffeine coursing through our veins, we spun a plan. Someday we would drive around the country on back roads and eat pie. It would be the Great American Pie Expedition. Their lives, alas, are now regulated by jobs, but mine grows freer and more unruly with each passing year, so Tazzie and I had undertaken the pie expedition. We were on its final leg; during the previous summer and autumn, we had been eating pie—some good, some bad, some indifferent—through Pennsylvania, southern New England, and on into Maine, where the blueberry and raspberry pies are glorious. Now we were heading west, through the cream-pie belt.

Before setting out on the initial trip that summer, I had wanted to establish some sort of standard for the expedition—a federal benchmark. I was in Washington, D.C., where I live for part of the year, so my husband and I went to lunch at the United States Senate Family Dining Room, which prides itself on its pie—particularly its pecan pie—nearly as much as it does on its bean soup. It seemed our patriotic duty to order, in addition to the pecan, the Senate Apple Pie, and only fair to add the special of the day, Senate Chocolate Mousse Pie. The last-named turned out to be a triumph of the imitation-food industry, and the apple pie was more government issue than federal benchmark. Both it and the pecan had been warmed up, though not, thank heaven, in a microwave oven. The microwave oven is one of the worst enemies that pie has in restaurant kitchens these days. I remember especially a piece of raspberry pie in Maine that had been microwaved to death. The hot raspberries, all integrity gone, oozed sadly out of the crust, and the pie had a faintly bitter aftertaste. Freshly made, warm pie is one of life's better things, but after it cools it should be allowed to grow old gently rather than brought back to an unnatural warmth. Our government-issue apple pie was no worse than many apple pies I was to sample on the road, but certainly no better. It contained too much cornstarch filling and too few apples, and those present were flaccid and tasteless. The pecan was a marginal improvement, but so cloying that it would have benefited from the quarter cup of brandy a friend of mine always adds to pecan-pie filling to cut its sweetness.

A few days later, Tazzie and I headed west and north in search of better pie. In western Maryland, on U.S. 40 at Sideling Hill, we drove through a magnificent roadcut, which is sometimes shown in geology textbooks; it is a place where a great part of the earth's history is laid out. There are signs along the roadside: "NO PARKING. NO CLIMBING." No gawking, no looking, don't be interested in what there is to see. When I had driven through there with Brian and Liddy on our trip a few years back, the State of Maryland wasn't so prohibitive, and we had parked, climbed, gawked.

Tazzie and I followed U.S. 40 north into Pennsylvania, stopping by the bank of a stream for Tazzie to run and for me to have a cup of coffee from the thermos. It was June, and the cow parsnip would soon be in bloom; daisies and buttercups already were. Wild geranium was everywhere. A blackbird, pert and sweet-voiced, was singing. "There's a cabbage white," I told Tazzie, "and there's a tiger swallowtail." I was trying to interest her in butterflies, but she paid them no attention whatever.

I sat sipping my coffee and thinking about pie. Meat pies, according to the *Columbia Encyclopædia*, have been around since the days of the Romans, but no one mentions fruit pies—or any dessert pies—until the fifteenth century. Even after that, no one ate very much pie until the New World was settled. Americans were the first to understand what pie could be. For instance, the English had been making what they called pompion by cutting a hole in the side of a pumpkin, extracting the seeds and the filaments, stuffing the cavity with apples, and baking the whole. New Englanders improved on this, combining the apples and pumpkin and putting them in a proper pastry. Then they eliminated the apples and added milk, eggs, spices, and molasses to the mashed, stewed pumpkin. Molasses, the colonists' substitute for sugar, was so important for pies that on several occasions a New England town put off its celebration of Thanksgiving for a week or more in anticipation of a shipment of molasses from the West Indies.

Americans discovered that other tasty pies could be made from materials at hand. A mock-cherry pie could be made with Cape Cod cranberries (spiced with raisins from Spain). Vermont pie was made with apples—a fruit successfully transplanted from England—and syrup cooked down from the sap of maple trees. Pie has never been more loved than in nineteenth-century America, where it was not simply dessert but also a normal part of breakfast. The food writer Evan Jones quotes a contemporary observer as noting that in northern New England "all the hill and country towns were full of women who would be mortified if visitors caught them without pie in the house," and that the absence of pie at breakfast "was more noticeable than the scarcity of the Bible." I knew a farmer in Iowa who died at the age of ninety-three a contented man, for he had eaten pie at breakfast every day of his life.

SUE HUBBELL

Herrings Family Restaurant is on U.S. 40—on the eastern outskirts of Uniontown, Pennsylvania. It is called a family restaurant not because it is okay to wear bluejeans there or because the food is moderately priced (although both these things are true) but because the Herring family runs it. I sat in a rosy orange plastic booth and had a home-style platter of beef stew, and after eating it I hoped that the pies would be as good. There was a choice: banana, coconut, peanut-butter, lemon, cherry-cream, blackberry, apple, cherry, raisin, and peach. A friend had recommended the banana pie, so I ordered that, but I couldn't resist ordering blackberry, too. The waitress looked amused. "I never had anyone order two pieces of pie before," she said. The crust on both pies was superb. The banana was still a little warm, and the filling spilled onto the plate. The bananas had been mashed before cooking, so there was pulp throughout instead of discrete slices. It was a good pie, but the blackberry was even better—tart and beautifully seedy. It was dense with fruit and had just as much filler as was needed, no more. The maker of the pies was pointed out to me—a slim, bespectacled young man. He blushed when I congratulated him on his pie-baking abilities, and told me that he sometimes made the pies but his relatives also made them. "The whole family works at the business," he said. "Grandpap started it. We all work in it." He pointed to the waitress. "She's my cousin."

"We make pies every day," his cousin said as I paid my check. "Doughnuts, too. And bread. And we eat all of them, but we stay skinny."

In the days that followed, I remembered fondly the pie at Herrings, for I had pies so middling as to be not worth reporting and one that was bad enough to be notable. I would not have been cross about this particular peach pie if pie had not been so boasted of on the restaurant's menu, and if it had not been peach season. I had asked for the apricot pie, which I'd heard good things about, but apricot was not available the day I was there, and since it was the height of the peach harvest I didn't think I could go wrong ordering peach pie. The crust was tasteless and thick and rested heavily in the stomach. The cook had been stingy with the peaches, which were suspended in a gluelike filling. I've had bad peach pie *out* of season—I remember especially one canned-peach pie in El Reno, Oklahoma—but during peach season it is hard not to make a good peach pie. Fresh peaches should be sliced and sweetened with enough white or brown sugar to make them pleasant to the taste. Egg yolk mixed with flour is a good thickener for peach pie, and after the thickener, sugar, and peaches have been lightly combined they should be heaped up roundly in an unbaked shell, topped with more crust, and baked. Toward the end of the baking, the pie should be removed briefly from the oven and the top crust brushed with a bit of egg white and sprinkled with sugar and cinnamon; the pie is then returned to the oven to brown prettily. That's all there is to it.

Of course, the underpinning of that pie—the crust—should be adequate. A New York man I know who thinks about pie a great deal says that pie judgments are sexually dimorphic. He believes that women judge a pie by its crust, men by its filling. That's so, says my friend Abby, because women know that the crust is the hardest part to make. That's *not* so, says my friend Linda, who loves both parts too much but eats only the filling, in the vain hope that most of the calories have settled in the crust. The truth is that a good pie can be ruined by a bad crust but a good crust cannot save a bad pie. A crust can be only so good, anyway. It should show off the pie and not call a lot of attention to itself. On my pie expedition I often found that restaurant cooks used frozen pie shells rather than making crusts themselves. In many cases, that was probably a mercy, because if left to their own devices they would make worse. Once, after I had eaten a memorably bad piece of cherry pie in a café in West Virginia the cook and the waitress sat down in my booth to talk. I asked the cook how she made the crust, which was thick and soggy. She told me that it was made of Crisco, flour, and milk. I've had a few bad milk crusts, and I don't use milk myself. Seldom has there been a week in my adult life when I haven't made a pie, and the following is the crust I use more often than any other. It is easy to make, crisp in a way that flakier crusts can't be, and shows off many pie fillings to advantage.

DOUGH FOR A TWO-CRUST PIE
2 C. flour
1 tsp. salt
½ C. cooking oil
¼ C. water

Mix flour and salt in a bowl and make a hollow in center of the mixture. Quickly blend water and oil together in a cup and pour into hollow in flour-salt mixture. Combine ingredients quickly and lightly with a few strokes of a spoon. Pinch dough in half and roll out each half between two sheets of waxed paper. Peel off top layer of paper. Invert dough into piepan and remove second layer of waxed paper.

On several occasions in the course of my expedition I stopped at the Mount Nittany Inn, which is just north of Tusseyville on Pennsylvania Route 144, hoping to have its special peanut-butter pie. A man whose pie sense I trust had urged me to try it. "Just don't think about it too much," he said. "Once you get over the idea of it, it's really terrific." But each time I went there the peanut-butter pie was not available, for one reason or another. The inn's proprietor agreed that it was a wonderful pie; his wife made it, he told me, but he didn't know how. "She makes the peanut butter all creamy, and then she freezes it," he said. "But I don't know what all she puts into it. A lot of stuff. Cheese, maybe?" I do hope not. In an effort not to think about *that*, I finally ordered, on my third stop, the inn's walnut pie, which was very good indeed. The crust was perfect, and the filling was similar to that of a good pecan pie, with black walnuts substituted for the pecans. The whole was topped with cream, whipped gently. It was satisfying, but not satisfying enough to drive the peanut-butter pie from my mind. I wish I could stop thinking about it.

I also failed to have the famous sour-cream raisin pie at the Potato City Motor Inn, in Potter County, Pennsylvania. That harsh, empty country, on the northern edge of the state, comes into its own in winter, during hunting and snow seasons. In daisy time, the Potato City Motor Inn is a quiet, unvisited place and serves up no sour-cream raisin pie. I asked the waitress which of the three pies available that day I should order. "None of 'em," she said. The Potato City Motor Inn was originally built as a place for potato growers to meet. Dr. E. L. Nixon, uncle to Himself, was involved in the beginnings of Potato City, where he crossbred potatoes and developed new strains. The cavernous dining rooms of the inn were empty when I was there. I would have liked at least to see the nearby ice mine, "a deep mountainside shaft" that "for no apparent reason forms heavy ice beginning in the spring," according to my Pennsylvania road map, which adds, "In winter the ice disappears." But it was closed. Potato County was in commercial diapause, so we drove on.

I stopped to feed Tazzie in a spruce woods, where bracken grew among the trees. While she ate, I watched clouds of blue butterflies—spring azures—squeezing into nearly opened blackberry blossoms to lay their eggs. We drove east on U.S. 6, a gentle road that wound through small towns filled with well-kept late-nineteenth-century houses surrounded by poppies. In between towns, wild blue phlox and buttercups were in bloom. I heard an Eastern wood pewee call *pee-a-weeee, peee-a-wee*. But the pie was indifferent until the X-Trail Restaurant, in Mansfield, at the intersection of U.S. 6 and Business Route U.S. 15. The X-Trail is a cheerful, unpretentious restaurant, painted blue, with crisp blue-and-white curtains at mullioned windows. It serves good pie baked in deep-dish piepans, which is just the way pies should be baked. (These pans are hard to find—something I discovered while shopping for one as a birthday present for Michael, my stepson, who has become interested in piemaking. But a few stores still stock them, and any serious piemaker should have one.) The X-Trail's coconut-cream pie was cool enough to cut, so I ordered some. It was a delicate pie, a handsome pie—monochrome, with white meringue floating on a cream white filling whose very pallor guaranteed that no packaged mix had been used. The crust was crisp, clean, distinct. The waitress told me that it was the restaurant's most favored pie—even more popular than the black-raspberry pie, which I also sampled, and which was stuffed with juicy berries.

I asked at each of my stops which pie was the most popular, and usually it was the coconut-cream. Coconut-cream is a good year-round pie. My friend Charlie, a Republican, is a pie conservative, and he doesn't believe in cream pies. The only real pies, he says, are of berries or other fruit, but he thinks that no one makes even fruit pies very well anymore. "You're not going to find good apple pie anywhere," he told me. I hoped he was wrong, but as the summer progressed it seemed possible that he might not be. When in doubt, I always ordered apple pie, and it was almost always as bad as the government-issue pie I'd had back in Washington—or worse. With such a standard, small wonder that gardeners with a lot of green tomatoes on their hands at the end of the season and with frost imminent tell us that green-tomato pie is just as good as apple any day.

Charlie wouldn't have approved of the chocolate-meringue pie I had at the Hotel Wyalusing, in Wyalusing, Pennsylvania, just off U.S. 6, by the Susquehanna River. I did. It was made of bittersweet chocolate so rich that the memory of Droste chocolate apples came to me. The meringue was as brown as a toasted marshmallow and so flat and neat that it must have

The Great American Pie Expedition

been spread with a knife. I would have enjoyed staying at the Hotel Wyalusing. It was a pleasant place—a former stagecoach stop lovingly restored, its brick front cleaned and its gingerbread woodwork, balconies, and dormers painted pale olive and buff yellow—but I wanted to eat pie at a New Jersey diner, and I wanted to get on to New York City to pick up a sweet-potato pie from Hugh Nelms, the president and corn-breadist of Hoecake International, so I drove on.

I discovered on my pie expedition that making a schedule was often a mistake. Certainly it was this time, because twenty-five miles west of New York City a tractor-trailer rig had tumbled over, and as a result all of northern New Jersey was in gridlock. I coasted, usually in neutral, across the state, reading the *Times* from page one through the classifieds, seldom passing the one-mile-per-hour mark. Pie at a diner was the first part of the plan to be abandoned, but I did want that sweet-potato pie. Abby had been telling me about it for years. She said that Nelms sells his pie every Sunday in fine weather at the open-air market at Seventy-seventh Street and Columbus Avenue. It is, she told me, a beautiful pie to look at, and a pie so tasty that she can't resist buying it in the six-inch-wide, hand-held version and eating it on the spot.

I tried to break away to the north, thinking I might come into the city from a different direction, but there had been a two-car collision somewhere in Connecticut, and that made the northern roads sluggish, too. Pie thoughts faded, and all I could think of was escape. I fled to the wilds of western Massachusetts and took stock. Sweet-potato pie would have to be added to my list of pies not sampled: Nittany peanut-butter, sour-cream raisin, apricot, sweet-potato. I don't like shoofly pie—a pie that has always reminded me of sweetened library paste—and wasn't even going to try eating that. I do like key-lime, but Florida was not on my route. Nor was Colorado, though a friend had recommended the black-bottom pie in

Denver. Nor Wisconsin, where there was said to be a strawberry pie so good that it was impossible to eat only one piece.

The next day was better. It was a golden, sunny morning, and beside the Connecticut River at Turners Falls, Massachusetts, I stopped in a park filled with bluebells to give Tazzie a short run. A friendly woman was walking her dog, who was part hound, part black Lab, and Tazzie fell in love with him. While our dogs played, the woman and I talked, and she told me about the Shady Glen, a restaurant on the main street of Turners Falls, where a friend of hers went once a day to eat chocolate-creme pie. The Shady Glen is a wide, squat, hospitable café. Inside are cheerful yellow booths and a wide selection of pies. I ordered the squash pie. It was my first of the season, nicely spiced with nutmeg and set off by a firm, crisp crust.

Diner chic is spreading, but it has not yet come to the Miss Florence Diner, on State Route 9, in Florence, Massachusetts, a vestigial outcropping of the more uptown-looking Alexander's Taproom. Miss Florence appears to have been there a long time and looks as though it would outlast Alexander's. It serves a thick, sincere apple pie—not an exceptional pie but a good pie. The apple pie at Allen Brothers Farm Market, on U.S. 5, in Westminster, Vermont, *is* exceptional. The indoor farm stand smelled of good things from the bakery the day I was there, but the fragrance of fresh apple pies dominated all. ("Thy breath is like the steame of apple-pyes," Robert Greene wrote in 1589.) I bought a whole one and a slab of Vermont cheddar to go with it, and then I drove down to Boston, where I was going to meet my husband and stay with Brian and Liddy, who live there.

It was an apple pie almost beyond praise. "This pie is good enough for breakfast," Liddy said reverently. The crust set it off well, and the apples—Lodis, in this case—were superb. Some of their skins had escaped the peeler and were in the pie, an addition that the four of us liked, but when I talked later to

Alice Porter, the Allens' baker, she was a bit embarrassed about them. Tim Allen, a second-generation apple grower, says Northern Spies make the best apple pie, and that's the variety Mrs. Porter uses when they're in season. When they're not, she uses Lodis or Cortlands.

ALLENS' PIECRUST
2 C. flour
¾ C. shortening
1½ tsp. sugar
1 tsp. salt
1 egg
1 tbsp. vinegar
¼ C. water

Mix dry ingredients in a bowl and cut in shortening with a pastry blender. In a separate bowl beat together egg, vinegar, and water. Mix with dry ingredients and refrigerate dough for at least two hours before rolling out. Makes a 9-inch two-crust pie.

ALLENS' APPLE-PIE FILLING
4 C. sliced apples
½ to ¾ C. sugar, depending upon tartness of apples
½ tsp. cinnamon
¼ tsp. nutmeg
2 tbsp. flour

Mix flour, sugar, and spices. Add to apples and mix lightly. Pour into unbaked pie shell. Dot with butter. Cover with top crust. Brush top with whole egg beaten with a little milk. Bake at 325° for 35 to 40 minutes or until browned.

Charlie, there's your pie.

In Maine, I settled into a routine of raspberry and blueberry pie, and each one was better than the one before it, which makes them very difficult to write about. The raspberry I remember with the greatest fondness is one I bought at the Village One Stop, in Lovell, in the western part of the state. One stop indeed: liquor, worms, gas, grain, groceries, lunch counter. A sign over the lunch counter, hand-lettered, said "PLEASE EAT OR WE'LL BOTH STARVE." I asked the young woman behind the counter about pie. She pointed to a man in a plaid flannel shirt sitting on a stool at the far end. "Talk to him—he just finished baking them."

"You have pie?" I asked him.

"Yep."

"What kind?"

"Pineapple. Blackberry. Raspberry."

"Raspberry sounds good."

Silence.

"May I have some?"

"Well . . . They're awful warm to cut. You want a whole pie?"

"Please."

The whole pie was five dollars and eighty-five cents. I bought it and a copy of the *Boston Globe*. The man smiled, faintly, and said, "Have a nice one. I mean a day, I guess. Have a nice day."

The warm pie filled the car with its fragrance. Tazzie and I drove north on State Route 5. The bracken at the roadside had the hint of bronze that says fall is on its way. The air and the sunlight agreed. We threaded in and out among logging trucks and finally pulled in at a rest stop, beside a lake where a loon was calling. A hand-carved sign on a post said that the spot was not maintained by any government agency and asked me to take away my trash. A reasonable request. The entreaty was signed by E. Littlefield—presumably the person who had mowed the grass so neatly and painted the blue picnic table, which was sheltered by a maple tree. Tazzie checked out the lakeshore, and I joined her, balancing on the rocks there. The late-morning sun made the wavelets glisten, their shimmer set off by the dark greens of spruce, fir, and pine that ringed the lake. I spread my *Globe* out on the picnic table, poured coffee from my thermos, and took out the pie. It was nicely browned and had little slash marks in the crust to let the steam escape. The edges were lovingly crimped, and the crust broke apart under my fork in delicious shattery flakes. The filling was no sweeter than it needed to be. It tasted of fresh raspberries and summer sunshine. I read my *Globe* and listened to the loon and ate pie. Thank you, E. Littlefield.

We resumed our drive north, but the blue sky and the sunshine prompted us to stop beside the Sunday River. Tazzie made friends with a man in sweatshirt and jeans who was peering at trees in a puzzled way. He asked if I had a field guide to trees with me. I did. We talked trees, and he told me that he had just begun learning to identify trees and birds and was driving around the country trying to learn the names of all he saw. Splendid man. I asked him if he'd like five-sixths of a really good raspberry pie. He said yes, and I gave it to him.

One afternoon, we stopped beside the Kennebec River, below the town of Skowhegan and its falls. Tazzie pounced on crickets, catching none. I watched damselflies. Clouds thickened and turned gray as they floated up from the southwest. Goldenrod and aster, the yellow and purple of summer's end, bloomed around us. I could hear a chain saw somewhere. All along the roads I'd been driving, I'd seen serious, multicord woodpiles. Winter is never far from the thoughts of people who live in north country.

Tazzie, who takes more interest in rocks than most dogs do, and certainly more than is good for her teeth, fetched numbers of them out of the Kennebec and laid them neatly on the bank, sometimes sticking her entire head underwater in a most undoglike fashion. We drove on through towns whose front lawns were bright with mountain ash, through Passadumkeag, through Mattawamkeag—which is to say that we were on our way to Aroostook, or The County, as Down Easters call it.

And there, in Houlton, I found the Elm Tree Diner, on the southwestern edge of town, on U.S. Alternate 2. It specialized in homemade pies, and it was a busy place, with table-filled additions that were signs of its commercial success. The blueberry pie had a thick but light crust. The berries were sparingly sweetened and little cooked; they maintained their integrity and berryish freshness so well that they brought back the memory of a sunny day when Brian, then twelve, and his father and I walked up Cadillac Mountain and stuffed ourselves with blueberries that we picked as we climbed.

I would have liked to buy pie in Wytopitlock, but there was none to be had, so I stopped in front of a boarded-up general store ("FOR SALE"), amid an unkempt patch of orange hawkweed, red clover, and pesky ripe burdock waiting for a chance to entangle itself in Tazzie's fur. There I ate a piece of French apple pie that I'd had the foresight to buy at the Elm Tree Diner. It is the same pie that in Pennsylvania is called Dutch apple pie—a single-crust apple pie topped with a crumbly mixture of brown sugar, flour, and butter (the "dowdy" of apple pandowdy). This French apple, however, was much better than any of its Pennsylvania Dutch relatives that I had sampled. It was generously filled with apples, and was pleasantly tart. It is always a mistake to sweeten these pies much, because the crumbly dowdy is sweet enough, and needs the sharp fruit contrast to be at its best.

Never play cards with a man named Doc, never eat at a place called Mom's, and never go to bed with anyone who has more troubles than you do, Nelson Algren advised. I discovered an exception in the case of a place called Mom's on U.S. 1, in Harrington, where I met Liddy and Brian for a traveling pie party. We ordered strawberry-rhubarb, apple, and blueberry pies and shared them. The strawberry-rhubarb was out of season but good nevertheless, and the apple was good, too, with a lovely ooze-browned back to its crust. The blueberry was the best I'd had so far. The crust was good—thick but not heavy—and the fresh blueberries that filled it were wild ones, sparkling and tangy. And then there was the blueberry pie we had at Duffy's, in East Orland, which was the last one I ate and the best one of all. The legend on the cover of Duffy's menu sounds a bit truculent:

>   Welcome to Duffy's
> We Here at Duffy's Are a
> Native Orientated Restaurant
>   We Aren't Fussy
> And We're Certainly Not Fancy.
>   If You Are,
> Ellsworth Is 12 Miles East
>   And
> Bucksport Is 7 Miles West.
>   Yours Truly,
>   Duffy

Even so, Duffy's is a welcoming sort of place, with geranium-filled window boxes. We ordered graham-cracker pie first. "You know what banana-cream pie is?" the waitress asked. "Well, it's like that, without the bananas." That's not quite accurate. It was more like a custard pie on a graham-cracker crust, topped with whipped cream and sprinkled with graham-cracker crumbs. Quite good, and much better than most custard pies, which are at best sweet, modest little things. But the three of us gave the blueberry pie gold stars all around. The crust had been pinched up into extreme points, and was delicious. The local blueberries were delicious, too, and their flavor was enhanced by the generous addition of cinnamon.

In season, though, all blueberry pies are good. A few years ago, I was visiting a non-baking friend, who asked me to make one. The oven wasn't working, but we thought we might be able to bake it on the covered outdoor grill, over a wood fire. There was no piepan, but perhaps I could make do with an aluminum cake pan. There was no rolling pin or waxed paper, so I used a water glass to roll out the pie dough between two pieces of brown paper cut from a grocery bag. There were, however, plenty of fresh wild blueberries, and there was sugar and cinnamon and a lemon, which I cut up and added to the filling. The pie that came off the grill had rather too thick a crust, and it had cooked unevenly, and it tasted of wood smoke, but it wasn't a *bad* pie; the blueberries were too good for that.

I had two other pies of note in Maine. The Farmington Diner, under the sign of fork and spoon transverse, was on State Route 4, on the south side of Farmington; it laid out logger-sized meals and good pie. The lunch special the day I was there was two thick pork chops cut from a very large pig, a platter of carrots, a platter of mashed potatoes and gravy, and a soup bowl full of applesauce. I made tiny inroads on the food, and realizing that I'd badly neglected lemon-meringue pie I ordered a wedge of it. It came accompanied by coffee in a sturdy mug that fitted amiably into my hand. My eyes told me that the pie filling was too yellow to be anything but a mix, but my taste buds said they didn't care: It was a good, classic diner pie—lots of loft to the meringue, and the crust a bit relaxed. Real lemon-meringue pie is too delicate and too ephemeral to be served in a diner, and too fussy to make for a diner cook to be happy with it.

The Milbridge House, in Milbridge, near the intersection of U.S. 1 and U.S. Alternate 1, serves a tasty and unusual pie. It is called Nantucket Cranberry Pie, and the recipe was brought up to Maine by Greg Charczuk, who owns the restaurant, when he moved from New Jersey. I copied it from the stained scrap of paper that his wife, Helen, uses when she makes it.

This is a fine pie, but, because the crust is cakelike and all on top, there will be pie conservatives who won't accept it. I will. If a creative and artful cook invents something and calls it pie, I'll call it pie. Let the crabbed formalists make their categories; I wouldn't like to miss something as lovely as this cranberry-walnut pie—an echo of the colonists' mock-cherry—or the banana-puddin' pie, which will make its appearance later.

NANTUCKET CRANBERRY PIE

3 C. fresh cranberries
Sugar to coat berries
½ C. chopped walnuts
1 C. flour
1¼ tsp. baking powder
⅛ tsp. salt
¼ C. shortening
1 C. sugar
1 egg
⅓ C. milk
½ tsp. vanilla
⅛ tsp. almond extract

Rinse cranberries, and dredge with sugar. Pour into greased 10-inch piepan, leaving excess sugar in bowl. Add chopped walnuts to cranberries. Cream shortening with 1 C. sugar. Add dry ingredients alternately with mixture of egg, milk, vanilla, and almond. Spoon and spread over cranberries and walnuts. Bake at 350° for 25 to 30 minutes.

My autumn was a busy one, and then the snows came, so it wasn't until the highways cleared in early springtime that Tazzie and I headed south from Washington on the pie roads.

Just south of Buffalo Gap, Virginia, on State Route 42, I came upon Our Place. It was part house, part restaurant, and was presided over by Betty Wade, a blond, comfortable woman with an easy smile. She showed me the back living room, where her daughters used to play until they were old enough to help out in the restaurant. Thirteen years ago, she told me, she and her husband bought the restaurant, and she started cooking there using *The Better Homes and Gardens Cook Book*, just as she did at home. "I wasn't sure folks would like my cooking," she said. "But they did. I didn't have to change anything." It seemed like the right kind of place to order butterscotch pie. I did, and watched the waitress slip quarters into the Wurlitzer.

> Shoe string, you ain't got no money.
> Shoe string, you cain't hang around here.
> Shoe string, you got your hat on back'ards.

My pie arrived. It was a trim, neat, light brown wedge on a crisp crust, topped with daintily browned meringue. Home cooking: I am back in the 1940s, hungry as only an eight-year-old can be with supper still half an hour away. I am standing in front of the open door of a refrigerator—a big one, with gently sloping shoulders. Inside, on the shelf next to the ice-cube-tray compartment, is a row of tall stemmed dessert glasses, each one filled with its own golden dollop of butterscotch pudding. I count. There are five, and five of us will sit down to supper. There is no way I can winkle one out of the refrigerator without drawing down maternal wrath. I can't remember what happened after that. I can't even remember eating the butterscotch pudding after finishing my nice vegetables. All I can remember is the yearning. I'm glad to be an adult. As I leave, George Jones is wailing from the Wurlitzer:

> The last thing I gave her was the bird,
> And she returned the favor with a few selected words.

We drove on. State Route 42 is my kind of road, playing tag with a little river that glistened in the sunlight. Well-tended farms are tucked in among the hills. We stopped at a wayside, superior in every respect to interstate rest stops, and Tazzie ran about, sampling the river, snuffling expectantly at the newly softened ground. A nuthatch in a sycamore at the river edge called *whoink . . . whoink* as he surveyed the bark for edibles. I leaned against the same tree and soaked up the wan sunshine, glad to be on the road, glad to be driving south to meet the springtime.

Pie called, and we got back in the car and went west on U.S. 60, a more peaceable road than the distant but parallel interstate. In Grayson, Kentucky, on Main Street, I stopped at the City Café, in accordance with pie rules I have formulated over the years for the Middle West. Rule 1: Pie is good in 85 percent of the eating establishments that are between two other buildings. Rule 2: Good pie may often be had near places where meadowlarks sing. If you follow these two rules in the Middle West you will find yourself at the City Café or its like. A restaurant like this makes it unnecessary for a town to have anything but a bland local newspaper. Nearly all the citizens gather there early in the morning for breakfast or coffee and the exchange of news and gossip. By 8:30 A.M., whatever has happened during the past twenty-four hours had been talked over. These cafés have a pressed-metal ceiling, sometimes covered with insulating panels, and dark imitation-wood paneling on the walls. Oil paintings by local artists are often displayed there for sale. The doors open early—by five or five-thirty. Fresh-baked pie is ready by nine-thirty or ten. The coffee drinkers are replaced by the dinner crowd not long after (away from cities, dinner is the noontime meal), and by two the pie is gone and the restaurant is closed for the day.

There was a Mountain Dew sign in front of the City Café. A poster in the window promoted a local wrestling match. Inside were an imitation-brick carpet and imitation-woodgrain tables. Everything was imitation except the food. I ordered pecan pie. It was exceptional—solid, with good texture. The filling was rich and eggy tasting but not overpoweringly sweet. The pecans on top were chopped, glazed, brown. Wilma Berry, a big woman with a pleasant face and a curly hairdo, was the waitress and owner of the café. She was happy to share her recipe, adapted from *What's Cooking in Kentucky:*

PALESTENE LAYNE'S PECAN PIE

1 tsp. vanilla
3 eggs, slightly beaten
1 C. corn syrup, light or dark
½ tsp. salt
½ C. white sugar
½ C. brown sugar
2 C. coarsely chopped pecans
1 unbaked 9-inch deep-dish pie shell

Blend well, but do not overbeat, vanilla, eggs, corn syrup, salt, sugar. Stir in pecans. Pour into pie shell. Bake approximately 50 minutes in preheated 350° oven, or until knife comes out clean.

The Beaver Dam Café was on the main street of Beaver Dam, Kentucky, just off U.S. 231, and was nestled between the Style Shop and Catalyst Management, Ltd. The sign outside said "HOME OF GOOD FOOD." Inside, imitation-needlepoint placemats said:

Cherish Yesterday
Dream of Tomorrow
Live Today

Nice, I thought; the Beaver Dam Café would have good pie. The possibilities, according to the menu, were interesting: cherry, pinto-bean, pecan, banana-pudding, and (Sundays only) buttermilk. It wasn't Sunday, nor was the pinto-bean ready, so I asked the waitress what I should order. "We sure do brag on our banana-puddin' pie," she said. I ordered banana-puddin' pie. It was fresh from the kitchen, chock-full of bananas, and so warm and relaxed that it had to be served in a bowl. The manager—a birdlike, wary little woman—said she'd be happy to give me the recipe but not her name. "What if someone tried to make it and didn't like it?" she asked. I am

The Great American Pie Expedition

able to reassure her. Over the past months, my stepson has been committing pie courtship. He and his friend Barbara have been having pie dates, making pies from the recipes I brought back from my expedition. They tell me that the pecan and apple pies were the best all around and Shaker lemon the prettiest and most flavorful, but that the banana-puddin' pie was the most fun.

BEAVER DAM CAFÉ
BANANA-PUDDIN' PIE
4 eggs, separated
3½ C. milk
¼ C. flour
¼ tsp. salt
1¾ C. sugar
4 tbsp. butter or margarine
2½ tsp. vanilla
Vanilla wafers
Bananas

Beat egg yolks with milk and add to mixture of flour, salt, and 1½ C. sugar. Cook in a double boiler over medium heat until thickened. Remove from heat and stir in butter and vanilla. In the bottom of a loaf pan put a layer of vanilla wafers and then a layer of sliced bananas. Pour one half of the custard over them. Repeat. Whip egg whites until stiff, adding ¼ C. sugar near the end. Spoon meringue on top of pie and bake in 425° oven until meringue is slightly browned.

A generation ago in the Ozarks, where I farm, pies also served romance—and an unlikely adjunct, school finance—in what were known as pie suppers. Back in those days of one-room schools in rural areas with a poor tax base, pie suppers were an annual autumn event. Young women would bake the best, the prettiest, the fanciest pies they could and take them to the school in the evening for young men to bid on. The top bidder would earn not only the pie but the right to eat it with its baker. The money from the auction funded the school. The young women would try to mark their pies in such a way that certain young men would recognize them, and Ozark folklore is full of stories about men who created emotional havoc by bidding—perhaps mistakenly, perhaps not—on the "wrong" pie.

In addition to Pie Rules 1 and 2, there is another, which applies to the entire country. Rule 3: Never eat pie within one mile of an interstate highway. This rule eliminates pie in most fast-food restaurants and in most truck stops, which are usually also franchises these days. I once violated Rule 3 and had a disappointing piece of gooseberry pie at a much recommended truck stop just off Interstate 70, between Kansas City and St. Louis. But now I thought I had better check out the small, independent truck stops, so I visited the Wyatt Junction Truck Stop, just west of the Mississippi River, on U.S. 60. A sign on the wall said:

WELCOME TO
WYATT JUNCTION TRUCK STOP
TRUCKER'S NOTICE
ALL COFFEE FREE WITH
THE PURCHASE OF DIESEL FUEL
THANK YOU FOR COMING
HAVE
A
NICE
DAY

I ordered the dinner special: chicken-fried steak, fried bread, deep-fried okra, French-fried potatoes, and a tossed salad made almost entirely of fried bacon. A trucker in a black leather jacket came in looking enormously pleased and announced to no one in particular that he'd passed his sweetie about two hundred miles back and left her behind. He hoped she wasn't frosted. He put a quarter in the jukebox:

The Great American Pie Expedition

> The last thing I gave her was the bird,
> And she returned the favor with a few selected words. . . .
> Then left two streaks of Firestone smokin' on the street.

Sweetie pulled up and climbed out of her own eighteen-wheeler. She sauntered in. She had a fluffy shock of black hair and was wearing tight jeans, high-heeled red shoes, and a black leather jacket to match his. She glared at him and ordered Royal Crown Cola and apple pie. If she could, I could. I told the waitress to hold the R.C. but lemme have some pie. It was very like that served in the United States Senate Family Dining Room.

I was just a bit bilious by the time that I got to my farm, in southern Missouri, so I spent a couple of days there sucking on soda crackers and letting Tazzie visit her favorite places down by the river. I had heard about Opal Wheeler's Pie Factory, on U.S. 63, south of West Plains, but I'd never been there, and I enlisted the help of my friend Nancy for a pie foray. Nancy is a little bitty skinny woman who runs a health-food store and talks a lot about bean sprouts. She is always ready to try something new, however, and is a woman of considerable enthusiasm. We drove down U.S. 63 and found the Pie Factory, a cheery, small ten-sided building with white walls and red-trimmed windows. It would have been easy to miss, because it sits back from the road and has only a small, hand-painted sign to call attention to itself. Opal Wheeler is a grandmotherly-looking woman with a warm smile. She had been in the restaurant business for years before she and her sister drove by the newly constructed building in 1985. Her sister, who understood Opal's love of baking, pointed to it and said, "Opal, wouldn't that make a cute little pie factory?" Opal agreed, rented the place, put up the sign, and set to work. She starts rolling out pie dough on the counter in the center of the building each morning at about five-thirty. She is usually done baking by eleven, and the rest of the day she sells pies—sometimes whole, sometimes by the slice with coffee.

The day Nancy and I were there she had apple, apple-raisin, raisin, pecan, cherry, apricot, blueberry, pineapple-cream, peach-cream, cherry-cream, icebox mixed-fruit, strawberry-rhubarb, gooseberry, lemon-meringue, chocolate, coconut, banana, chocolate-delight (chocolate, pecans, cream cheese, single crust, whipped topping), lemon-delight, pecan-delight, and strawberry. I ordered a cup of coffee and told Nancy she could order whatever she wanted if she would talk to me about it. She began with a wedge each of apricot, apple-raisin, cherry, icebox mixed-fruit, and strawberry. And a cup of tea. "Oh, look," she said. "The crusts are sprinkled with sugar and browned. Pretty. . . . This apricot is too mooshy. . . . The apple-raisin is superior, though. I think it's the best one. . . . No, the cherry is the best—tart, lots of cherries, not much gooey filling. . . . No, maybe the mixed-fruit is my favorite. It's got a single crust, then a layer of cream cheese, then cherries with fresh pineapple. And this whipped topping! Wow! Can this lady make pies!"

"Anything else appeal to you?" I asked.

"Well, maybe I'll try a piece of lemon-meringue."

I watched in admiration as she ate that, too. Opal Wheeler freshened my coffee. I asked Nancy again which pie was best.

"How can I choose?" she asked. "Each one is a jewel."

"Nancy," I said, "you run a health-food store."

"Yeah, isn't it wonderful? She can make good pies out of stuff like this. I mean, white flour, white sugar, solid shortening? Maybe she could make the crusts out of whole-wheat flour. Oh, well, it just shows you what a talented cook can do."

Nancy left with an entire chocolate-delight pie in a box. She was happy, in perfect health, looking not one ounce fatter. And I was happy. I had always hoped that pies were good for us—a hope that had been encouraged by an article in the *Weekly*

World News of October 27, 1987. I've long thought that there is a supermarket-tabloid headline designed to sucker every man, woman, or child at least once. Mine was "SNICKERS AND TWINKIES MAKE YOU HEALTHY, SAYS FOOD EXPERT," touting an article declaring that there was "more nutrition in a Snickers bar or a Twinkie than in an apple." Might I not therefore assume, after watching Nancy tuck into her pies, that apple pie was the healthiest way to eat an apple?

In western Missouri, the good pie places thin out. The town cafés have become Daylight Donut outlets, to the detriment of both pies and doughnuts, but just at the border of Oklahoma I had an unexpectedly dainty and tasty chocolate pie at the Corners Minimart Motel & Café, on U.S. 60. The chocolate filling rested on a flavorful crust and was topped with a perfect, delicate meringue. Outside, prickly pear grew on the sunny, south side of the restaurant, where there was a big white box with a sign on it that said:

CAUTION
BABY RATTLERS

Hints of the West.

Once I had crossed the Oklahoma line, I began seeing red-tailed hawks hunting high in the air above the road. It had been a long time since I'd heard a meadowlark sing. In one small town after another, waitresses in the cafés shook their heads when I asked for pie, and offered cobbler instead. "And mighty good cobbler it is, too," a customer informed me in the Hot Biscuit Café, in Vinita. But I drove on. There were sandburs at the rest stops, and Tazzie whimpered when she got them in her feet. The sun was warm. Spring had been in Oklahoma for some time. The sky opened up. I could see forever. The road threaded between hills covered with prairie grass, and someplace between Bartlesville and Ponca City I realized that I was in the West. I rolled down all the windows. *Ky-y-y-yr* screamed a red-tail overhead. I leaned out the window and the wind blew and tugged at my hair. "*Ky-y-yr*," I screamed back. Fun. Can't do that on the interstate. The road was empty, and I was in love with driving. Out of nowhere, an Oklahoma trooper came up behind me and pulled me over. He reminded me that the speed limit was still fifty-five off the interstate, and gave me a "courtesy" ticket. Nice young man. Troopers know pie, so I asked him where he went to get it. He blushed a little, took off his hat, scratched his head, and thought a while. Then, by way of explanation, he said, in his slow drawl, "Sorry, Ma'am, but you're in cobbler country now."

—The *New Yorker*, March 27, 1989.

The Vicksburg Ghost

> The human predicament is typically so complex
> that it is not altogether clear which lies are vital and
> what truths beg for discovery.
>
> —*Vital Lies, Simple Truths: The Psychology of
> Self-Deception*, BY DANIEL GOLEMAN, 1985

I guess most people found it hard to believe that Elvis Presley didn't die after all but instead is alive and well and shopping at Felpausch's Supermarket, in Vicksburg, Michigan. I know I did when I read about it in the *New York Times* last fall. The *Times* wasn't on record as saying, "THE KING LIVES," or anything like that, but it did report that a Vicksburg woman named Louise Welling had said she'd seen him the year before, in the supermarket's checkout line. Her sighting encouraged Elvins everywhere, many of whom believe that Presley faked his death. It also added an extra fillip to Elvismania, which is part nostalgia and part industry, the industry part consisting of the production of Elvis memorabilia, books, articles, tours, and prime-time TV "docudramas." Fans have made periodic demands for an Elvis postage stamp, and a multimedia musical—*Elvis: A Rockin' Remembrance*—had an Off-Broadway run in 1989.

Promotion was what made Elvis Presley. In 1977, the year of his death, his likeness was more widely reproduced than any other save that of Mickey Mouse, and it has been reported that the news of his demise was greeted by one cynic with the words "Good career move!" According to Albert Goldman, the biographer who tells this story, Presley was by then a porky, aging, drug-befuddled Las Vegas entertainer and was getting to be a hard personality to promote. The Presley image shorn of the troublesome real man was easier to market. For example, after the King's death, Presley's manager, Colonel Thomas A. Parker, contracted with a vineyard in Paw Paw, Michigan—a town not far from Vicksburg—to produce a wine called Always Elvis. Its label bears a head shot of the entertainer, in a high-collared spangled white shirt, singing into a hand-held microphone. Colonel Parker's own four-stanza poem appears on the back of the bottle. Goldman has computed that the poem earned Parker twenty-eight thousand dollars in royalties, "making him, line for line, the best-paid poet in the world." Although the wine is no longer produced, I was able to find a dusty old bottle in my local liquor store. In the interests of journalism, I sampled it. It was an adequate companion to the poem, which closes with the couplet

> We will play your songs from day to day
> For you really never went away.

In its year-end double issue, *People* ran a story featuring recent photographs of Elvis purportedly taken by readers around the country, each picture as vague and tantalizing as snapshots of the Loch Ness monster. While debate mounted over whether or not Elvis Presley was still alive, I got stuck back there in the part of the *Times* story which said that he was shopping at Felpausch's. By the latter part of the 1950s, when Elvis arrived to sweep away the dreariness of the Eisenhower years, I was too old to respond to the Dionysian sexual appeal that he had for his teenage maenads; consequently, I was also unmoved by retro-Elvis. But I did grow up near Vicksburg. My family lived in Kalamazoo, a bigger town (in which Elvis was also said to have appeared) twelve miles to the north, and we spent our summers at a lake near Vicksburg. My widowed mother now lives at the lake year-round, and when I visit her I often shop at Felpausch's myself. I know Vicksburg tolerably well, so when I read the account in the *Times* I strongly suspected that the reporter had been snookered by a group of the guys over at Mar-Jo's Café, on Main Street, half a block from Felpausch's, which is on Prairie Street, the town's other commercial thoroughfare. Last June, while I was visiting my mother, I decided to drive into Vicksburg and find out what I could about the Elvis Presley story.

Vicksburg is a pretty village of two thousand people, more or less. A hundred and fifty years ago, when it was first settled by white people, the land was prairie and oak forest. James Fenimore Cooper, who lived for a time in the nearby town of Schoolcraft, wrote about the area in his book *Oak Openings*. It is in southern Michigan, where the winters are long and gray, and even the earliest settlers complained of the ferocity of the summertime mosquitoes. Vicksburg's one-block commercial section has been spruced up in recent years. There are beds of petunias at the curb edges, and new façades on the nineteenth-century buildings. The carefully maintained Victorian houses on the side streets are shaded by maples big enough to make you think elm. A paper mill, built near a dam that the eponymous John Vickers constructed on Portage Creek for his flour mill, has long provided employment for the local people, but today the village has become something of a bedroom community for commuters to Kalamazoo. Still, it seems very like the place I knew when I used to come to band concerts on Wednesday evenings at the corner of Main and Prairie, during the summers of the 1930s and '40s. The band concerts are a thing of the past, but there are other homegrown entertainments, such as one going on the week I was there—the annual Vicksburg Old Car Festival, which is run by Skip Knowles, a local insurance man. The festival has a fifties theme, and last year, inspired by the commotion that Louise Welling's sighting of Elvis had produced, Knowles added an Elvis-look-alike contest to the roster of events. Knowles has his office in a storefront on Main Street which used to be Matz's Confectionery, where I first discovered lime phosphates (known locally as "green rivers").

And the teenagers are still bored. While I was in the library going through back issues of local newspapers, two high-school girls introduced themselves to me, saying that they had lived in Vicksburg all their lives and would be happy to talk to me about it. I asked them what they thought about Elvis Presley. They smiled patronizingly and informed me that no one they knew paid any attention to him. "But *everything* just stands still in Vicksburg," one of them confided. "We go to Kalamazoo on Saturday nights. I can't wait to get out of here and go to college."

Mar-Jo's has stayed the same, too. It has been in the same place for forty years. It was named after Marge Leitner and her partner, Josephine, whose last name no one at the café can remember. It is your basic tan place: tan floor, tan walls, tan tables, tan counter. The sign taped to the cash register was new to me. It said:

>   THIS IS NOT
>   BURGER KING
>   YOU GET IT
>   MY WAY
>   OR YOU DON'T
>   GET IT
>   AT ALL

But the men having coffee together at the big round table near the front windows could have been the same ones sitting there the last time I was in, which was a couple of years ago.

"How's you-know-who?" gray crewcut asks feed-store cap. "Don't see her anymore."

The others guffaw, and one says, "He's taken her clothes."

"What clothes?" feed-store cap shoots back. A ripple of caffeine-fueled laughter circles the table.

The Vicksburg Ghost

Shirley White, a small, wiry woman, has been a waitress at Mar-Jo's for eleven years. Her hair is dark and tightly curled. She is efficient and cheerful. She knows virtually all her customers by name and how they like their coffee, and she banters with all of them. She gets to work at four-forty-five every morning, so she is usually way ahead of the best of the town wits, giving as good as she gets. The coffee-club boys once arranged the kind of prank on her that made me suspect them of the Elvis Presley caper. One of the regulars was a big man whom she could deftly unsettle with a clever phrase or two. His invariable riposte was a mumbled "Paybacks are hell." A few years ago, he was on vacation in Florida when her birthday came around, and she had nearly forgotten about him. Mar-Jo's was jammed that day, and no one would tell her why. "Just as I was busiest, this really big monkey walked in," she told me. "At least, it was a big guy dressed in a monkey costume, and he kept following me around, getting in my way. I was real embarrassed, and everyone kept laughing. Then a messenger handed me something called an Ape-O-Gram. It had just three words: 'Paybacks are hell.'"

Nearly all the coffee drinkers thought that the Elvis Presley sighting was as funny as the Ape-O-Gram, but no one would own up to having had a hand in making up the story. Louise Welling, it seemed, was a real person, and well known in town. She lived to the east, a few miles outside the village, they told me. "She's different, that's for sure," one of the coffee drinkers said. "No one believes her about Elvis Presley, but we all enjoyed it. Kind of put Vicksburg on the map. Isn't it funny? Elvis Presley wasn't even a very good singer. But I don't think Louise thinks it's funny." They referred me to a woman in town who knew Louise Welling better than they did and lived not far from her.

I went over to see the woman, who had an office in town, and talked to her with the understanding that her name would not be used. "Yes," she said. "I guess you could say that Louise is different. Her whole family is different, except for her husband, who works at General Motors. He's real quiet. But she's not crazy or anything. In fact, I think she's real bright. I don't know what to make of her claim that she saw Elvis Presley. She was a big Elvis fan from way back, but she doesn't bring him up or talk about this stuff unless someone asks her. She's a kind woman. She's reliable, too, and I wouldn't hesitate to call her if I had trouble. I'm afraid that after the story came out a lot of people played jokes on her. Made Elvis phone calls. Sent her Elvis letters. I'm pretty sure she's not in it for money. She just seems to think it's an interesting story, and it makes her mad when people don't believe her. Of course, none of us do. I don't know anyone in this town who thinks she really saw Elvis Presley. She was furious with the Vicksburg newspaper because they wouldn't run her story."

It seemed odd to me that the *Vicksburg Commercial* had not used Louise Welling's story—a story that had made the *New York Times*—so I called up Jackie Lawrence, the owner of the *Commercial*, and asked her to meet me for lunch at Mar-Jo's. Jackie Lawrence, a former nurse, is a big woman with curly brown hair, and she smiles a lot when she talks about Vicksburg, her adopted town. There are, she said, perhaps a dozen loyal Elvis fans in town—people who make pilgrimages to Graceland and would *like* to believe Louise Welling even if they don't.

We studied the daily specials, which were posted on the wall, and I decided to order Ken's Homemade Goulash. Next to the list of specials were snapshots of Ken Fowler, a cheerful young man with a fine brushy mustache, who bought Mar-Jo's two years ago and does a lot of the café's cooking. Shortly after he bought the place, he had a birthday, and the regulars, the waitresses, and Ken's wife conspired to bring in a belly dancer. The event was captured on film, and the posted snapshots show

Ken, in apparent embarrassment, on a chair in one corner of the café, surrounded by laughing customers as a woman in gold draperies writhes in front of him.

Jackie Lawrence told me that she remembered Louise Welling coming into the newspaper office, which is a few doors down from Mar-Jo's, in March 1988, six months after the sighting at Felpausch's. At the time of her visit, Mrs. Welling knew that her story would soon be printed nationally, in the *Weekly World News*—and so it was, three months later. (According to Jim Leggett, who is the dean of freelance tabloid photojournalists and once schemed to drill a hole in Howard Hughes's coffin in order to photograph his face, the *Weekly World News* is not exactly esteemed in the trade. "It prints the flotsam left by the better tabloids," he told me.) Mrs. Welling had wanted the *Commercial* to run her story first, Lawrence said. "She stood right by my desk, trying to tell me all about it. I said to her, 'I'm sorry, I don't have time for this,' and showed her out the door. And if she came in again, I'd say the same thing."

There was only one mention in the *Commercial* of the stir caused by Louise Welling's encounter with Elvis. The winner of Skip Knowles's 1988 Elvis-look-alike contest, a truck driver named Ray Kajkowski, came into the newspaper office a few days after the event to ask for prints of any pictures that might have been taken. While he was there, he kissed Jean Delahanty, one of the *Commercial*'s reporters, and she wrote a column about it, which concluded, "Some days are better than others!"

There is no chamber of commerce, as such, in Vicksburg. The town doesn't need one; it has Skip Knowles. I had telephoned Knowles before coming to Vicksburg. "Give me a jingle when you get in," he said. "Maybe we can do lunch." He is a handsome, trim, dark-haired man, and at our lunch a gold chain showed through the open collar of his shirt. There was another gold chain around his wrist. He was born in Atchison, Kansas, he told me, but spent his teenage years—from 1962 to 1968—near Detroit, where he developed a passion for cars and for cruising, that cool, arm-on-the-window, slow patrolling of city streets which was favored by the young in those days. His dark eyes sparkled at the memory.

"We had what we called the Woodward Timing Association," he said. "It was made up of the guys that cruised Woodward Avenue. The Elias Big Boy at Thirteen Mile Road and Woodward was the place we'd go. But you know how the grass is always greener somewhere else? Well, my ultimate dream was to cruise the Sunset Strip. It wasn't until I got married, in 1969, and went out to California that I got to do that. And I talked to those guys cruising the Strip, and you know what they told me? It was *their* dream to cruise Woodward." He shook his head and laughed. "My wife and I still cruise when we go to a city." He hoped the local people had got cruising down pat for this year's festival, he said, handing me a packet of publicity material and a schedule of festival events. "I had to *teach* them how to cruise last year, which was the first time we closed off the streets for it."

The second annual Elvis-look-alike contest would be held at 9:00 P.M. Saturday, over on Prairie Street, in the parking lot of the Filling Station, a fast-food restaurant across the street from Felpausch's. Skip Knowles knew a good thing when he had it. Before last summer, he said, the festival had been drawing several thousand people, but each year he had had more trouble getting good publicity. "I can't understand the way they handled the Elvis business over at Felpausch's," he told me. "They even refused an interview with the *New York Times*. But I decided to play it for whatever it was worth."

After the first Elvis-look-alike contest, Knowles received a lot of calls from Louise Welling, who wanted to talk about Elvis Presley with him. "I put her off," he said. "She's *really* different. I think she really believes Presley never died." He also received other phone calls and visits. When his secretary told him last fall that a reporter from the *Times* was in his outer office waiting to talk to him, he thought it was just a hoax—a joke like the ones dreamed up at Mar-Jo's. But when he came out the man introduced himself as the paper's Chicago bureau chief and interviewed him about the Elvis contest. Then a producer from Charles Kuralt's show, *Sunday Morning*, called and said he was interested in doing a segment for the show on the impact of the Elvis sighting in Vicksburg, and would anything be going on in Vicksburg around Thanksgiving time? "I told him, 'Look, I'll do *anything* to get you here,'" Knowles recalled. "'If you want me to rent Cadillac limos and parade them up and down Main Street for you to film, I'll get them.' But the TV people never came."

I decided that it was time to talk to Louise Welling herself. I couldn't make an appointment with her by telephone, because she had recently obtained an unlisted number, but one midweek morning I took a chance on finding her at home and drove out to see her. The Wellings lived in the country, in a modest split-level house on non-split-level terrain; this is the sandy, flat part of Michigan, too far south for the ice-age glaciers to have sculpted it. Mrs. Welling sometimes works as a babysitter, but this morning she was home, along with four of her five children—all of them grown—and Nathan, her four-year-old grandson. Mrs. Welling is a heavyset woman with closely cropped dark hair and a pleasant face. Her eyes stay sad when she smiles. She touched my arm frequently as we talked, and often interrupted herself to digress as she told me her story. She said that she grew up in Kalamazoo and for a time attended St. Mary's, a Catholic grammar school there. When she turned sixteen, she was given a special present—a ticket to a Presley concert in Detroit. "Somehow, the fellow who took tickets didn't take mine, so after the show I was able to move up, and I sat in front during the second," she said. "And then, toward the end, Elvis got down on his knee right in front of me and spread his arms wide open. Well, you can imagine what *that* would be like for a sixteen-year-old girl." Her voice trailed off, and she fell silent, smiling.

SUE HUBBELL

I asked her if she had continued to follow his career.

"When I got married, I started having children, and I never thought much about Elvis," she said. "After all, I had problems of my own." But then, in 1973, she saw a notice in a throwaway shopping newspaper from Galesburg, a nearby town, saying that Presley would be in Kalamazoo and, although he would not be performing, would stay at the Columbia Hotel there.

"I didn't try to get in touch with him," Mrs. Welling said, adding, with a womanly smile, "I had a husband, and you know how that is." Three years later, however, Presley appeared in concert in Kalamazoo, and she sent flowers to him at the Columbia Hotel, because she assumed that he would be staying there again. She went to the concert, too, and, as she remembers it, Elvis announced in the course of it that he had a relative living in Vicksburg. "He said he liked this area," she recalled. "Kalamazoo is a peaceful place. He'd like that. And I think he's living at the Columbia right now, under another name. But they won't admit it there. Every time I call, I get a runaround. You know what I think? I think he has become an undercover agent. He was interested in that sort of thing."

"What year was it that you saw him in concert in Detroit?" I asked. I had read somewhere that Presley had not started touring outside the South until 1956.

"Oh, I don't remember," Mrs. Welling said. "I'm fifty-one now, and I had just turned sixteen—you figure it out."

The arithmetic doesn't work out—nor, for someone who grew up in Kalamazoo, does the Columbia Hotel. The Columbia had its days of glory between the First World War and Prohibition, and it was growing seedy by the forties, when I used to drive by it on my way to school. Its decline continued after I left Kalamazoo, until—according to Dan Carter, one of the partners in a development company that remodeled the hotel to create an office complex called Columbia Plaza—it became "a fleabag flophouse and, for a while, a brothel." Carter also told me that in the mid-eighties a rumor arose that Elvis Presley was living there, behind the grand pink double doors on the mezzanine, which open into what was once a ballroom. The doors have been locked for years—the empty ballroom, its paint peeling, belongs to the man who owns Bimbo's Pizza, on the floor below—but that didn't deter Elvins here and abroad from making pilgrimages to Columbia Plaza. "You'd hear foreign voices out in the hallway almost every day," he said. "Then there was a visit from some people from Graceland—at least, they told us they were from Graceland, and they looked the part—who came by to see if we were making any money off this." They weren't, he said, and today the building's management denies that Elvis Presley, under any name, lives anywhere on the premises.

Mrs. Welling's next good look at Elvis Presley came at Felpausch's, in September 1987. There had been, she told me, earlier hints. In 1979, she had seen a man in the back of the county sheriff's car when the police came to her house to check on the family's dog, which had nipped a jogger. "The man in the back seat was all slouched down, and he didn't look well," she said. "I'm sure it was Elvis." A few years later, black limousines began to appear occasionally on the road where she lives. "Now, who around here would have a limo?" she asked. Then she began seeing a man she believes was Elvis in disguise. "He looked real fake," she recalled. "He was wearing new bib overalls, an Amish hat, and a beard that didn't look real. I talked to a woman who had seen the same man, and she said he sometimes wore a false nose. Now, why does he have to bother with disguises? Why couldn't he have said that he needed a rest, and gone off to some island to get better?"

A note of exasperation had crept into Mrs. Welling's voice. She showed me a cassette that she said contained a tape that Presley made after he was supposed to have died; in it, she said, he explained why he had faked his death. But when she played it the sound was blurred and rumbly, and I couldn't make out the words. The tape had been issued in 1988, to accompany a book by a woman—with whom Mrs. Welling had corresponded—who put forward the theory that the body buried as Presley's was not his own. The book and another by the same author, which Welling said was a fictional account of a rock star who fakes his death, were lovingly inscribed ("It's hard to take the heat") to Mrs. Welling.

Here is what Mrs. Welling said happened to her in September 1987. She had just been to eleven o'clock Sunday Mass at St. Martin's Church. With grandson Nathan, she stopped at Felpausch's to pick up a few groceries. Having just celebrated one publicly accepted miracle, she saw nothing strange in the private miracle at the supermarket.

"The store was just about deserted," she said. "There wasn't even anyone at the checkout register when I went in. But back in the aisles I felt and heard someone behind me. It must have been Elvis. I didn't turn around, though. And then, when I got up to the checkout, a girl was there waiting on Elvis. He seemed kind of nervous. He was wearing a white motorcycle suit and carrying a helmet. He bought something little—fuses, I think, not groceries. I was so startled I just looked at him. I knew it was Elvis. When you see someone, you know who he is. I didn't say anything, because I'm kind of shy and I don't speak to people unless they speak first. After I paid for the groceries, I went out to the parking lot, but no one was there."

I asked Mrs. Welling if she had told anyone at the time what she had seen. She replied that she had told no one except the author of the Elvis-isn't-dead book, who was "very supportive." After that, she and her daughter Linda started seeing Elvis in Kalamazoo—once at a Burger King, once at the Crossroad Shopping Mall, and once driving a red Ferrari. And she said that just recently, while she was babysitting and filling her time by listening to the police scanner, she heard a man's voice ask, "Can you give me a time for the return of Elvis?" and heard Presley reply, "I'm here now."

The Vicksburg Ghost

I asked her what her family thought about her experiences. Linda, a pale, blond woman who was sitting off to one side in a dining alcove smoking cigarettes while I talked to her mother, was obviously a believer, and occasionally she interjected reports of various Elvis contacts of her own. "But *my* mother thinks it's all nutty," Mrs. Welling said, laughing. "She says I should forget about it. My husband doesn't say much—he's real quiet—but he knows I'm not crazy."

It wasn't until the spring of 1988, Mrs. Welling said, that she started getting in touch with the media. She claims that she didn't bother talking to the people at the Vicksburg newspaper (although Jackie Lawrence remembers otherwise), because "it wasn't an important newspaper." Instead, she tried to tell her story to the *Kalamazoo Gazette* and people at the television station there. No one would take her seriously—except, of course, the author of the Elvis book. After Mrs. Welling had written to her and talked to her on the telephone, a writer for the *Weekly World News* phoned for an interview. Mrs. Welling asked him how he knew about her, but he declined to reveal his sources. In early May, the tabloid prepared the ground for Mrs. Welling's story by running one that took note of the rumor that Presley was living in Columbia Plaza, and gave Mrs. Welling's friend a nice plug for her book. Shortly after that, the syndicated columnist Bob Greene gave the rumor a push. By that time, the *Kalamazoo Gazette* realized that it could no longer ignore Mrs. Welling's phone calls, and in its May 15 issue Tom Haroldson, a staff writer, wrote a front-page story headlined "'ELVIS ALIVE' IN KALAMAZOO, SAY AREA WOMAN AND NEWS TABLOID." That was the beginning of Mrs. Welling's fame, but it was not until June 28 that the *Weekly World News* told her whole story. In thousands of supermarkets, the issue appeared with a big front-page picture of Mrs. Welling and a headline in type an inch and a half high proclaiming "I'VE SEEN ELVIS IN THE FLESH!" The story began to be picked up by newspapers around the country as a brightener to the increasingly

SUE HUBBELL

monotonous accounts of the pre-convention presidential campaigns. CBS investigated it for possible production on *Sixty Minutes*. Radio stations from coast to coast and as far away as Australia called to interview Louise Welling and anyone else they could find. Kalamazoo's mayor, Edward Annen, reacted to all this by announcing to a *Gazette* reporter, "I've told them that everyone knows this is where he lives and that they should send their residents here to spend tourist dollars to find him."

Funny signs sprouted throughout Kalamazoo and Vicksburg in places of commerce. A rival market of Felpausch's posted one that said "JIMMY HOFFA SHOPS HERE." A dentist boasted, "ELVIS HAS HIS TEETH CLEANED HERE." At Mar-Jo's, the sign read "ELVIS EATS OUR MEATLOAF." The folks at Felpausch's, however, were not amused. Cecil Bagwell, then the store's manager, told the *Gazette*, "The cashier who supposedly checked out Elvis that day cannot remember anything about it," and characterized Mrs. Welling as "an Elvis fanatic." Bagwell no longer works at Felpausch's, but I spoke with Jack Mayhew, the assistant manager, who scowled when I brought up the subject. "I won't comment," he said, adding, nonetheless, "We've never given the story to anyone, and we're not going to. All I'll say is that the woman is totally—" and he rotated an extended finger beside his head.

Before I left Mrs. Welling that morning, I asked her why she thought it was that *she* had seen Elvis, when others had not—did not even believe her.

"I don't know, but the Lord does," she answered. "I'm a religious woman, and when things like this happen—that we don't understand—it just proves that the Lord has a plan."

The next day, a friend who had heard about my investigations telephoned to tell me that there had been an Elvis sighting just a week or so earlier, in Kalamazoo, at the delivery bay of the Fader Construction Company, which is owned by her family. She hadn't seen the man herself, she said, but the women in the office had insisted that the truck driver making the delivery was Elvis Presley. I suspected that it might have been Ray Kajkowski, winner of the Elvis-look-alike contest and kisser of Jean Delahanty. This turned out to be true. On Friday evening, at a run-through for the Old Car Festival's cruising event, I was introduced to Kajkowski by Skip Knowles, and Kajkowski confirmed that he had made quite a stir while delivering a shipment of concrete forms to Fader. He gave me his card—he has apparently made a second career for himself as an Elvis impersonator at parties and night clubs—and then he whipped out a pair of mirrored sunglasses, put them on, and kissed me, too. "Young, old, fat, skinny, black, white, good-looking, not so good-looking, I kiss them all," he said. "I'm a pretty affectionate fellow. I was raised by a family that hugged a lot."

Ray Kajkowski lives in Gobles, not far from Vicksburg. At forty-one, he is thick-featured, a bit on the heavy side, and looks like—well, he looks like Elvis Presley. He has big sideburns and dyed black hair, which he wears in a pompadour. He went down to Graceland recently with his wife and his two teenage sons to study the Presley scene and recalls that while he was in the mansion's poolroom a couple came in and the wife took one look at him and collapsed on the floor in a faint.

"When I was growing up, I felt like an outsider," he told me. "I didn't think I was as good as other people, because my dad wasn't a doctor or a lawyer. We were just common folks. I knew about Elvis even when I was a little kid. I didn't pay much attention, though except that some of my buddies had pictures of Elvis, so we'd trade those to our older sisters and their friends for baseball cards." He laughed.

"I felt like we were invaded when the Beatles came over," he continued. By that time—1963—he was at Central High School in Kalamazoo, and had begun to appreciate Presley's music and to defend it against foreign stars. "I mean, Elvis was a small-town boy who made good. He was just ordinary, and, sure, he made some mistakes, just like me or you or any of us.

SUE HUBBELL

But he went from zero to sixty. He had charisma with a capital 'C,' and somehow people still know it."

After Presley's death, Kajkowski said, he felt sad and started reading about Elvis and studying his old movies. "Then, in September or October 1987, right around then, I was at a 1950s dance in Gobles. My hair was different then, and I had a beard, but there was a fifty-dollar prize for the best Elvis imitator. Fifty bucks sounded pretty good to me, and I watched this one guy do an imitation, and he didn't move or anything, and I thought to myself, I can do better than that, so I got up and entered and won, beard and all. After that, I shaved off my beard, dyed my hair, and started building my act. I do lip-synch to Elvis tapes. I've got three suits now, one black, one white, one blue. My wife does my setups for me and runs the strobe lights. Evenings when we don't have anything else to do, we sit around and make scarves for me to give away. I cut them, and she hems them. When I'm performing, I sweat real easy, and I mop off the sweat with the scarves and throw them out to the gals. They go crazy over them. And the gals proposition me. They don't make it easy. Sometimes they rub up against me, and when I kiss them they stick their tongues halfway down my throat. Once, I went over to shake the guys' hands, because I figured it was better to have them on my side. But one big guy wouldn't shake my hand, and later he came over and grabbed me like a grizzly bear and told me to quit it. 'You don't sound like Elvis Presley. You don't look like Elvis Presley. Stop it.' I told him, 'Hey, it's all lip-synch! It's just an act! It's entertainment!' But I try to keep it under control. My wife's the woman I have to go home with after the act."

I asked Kajkowski if he had ever been in Felpausch's. As a truck driver, he said, he had made deliveries there; occasionally, he even shopped there. But although he owned a motorcycle, he rarely drove it, and he never wore a white motorcycle suit.

I asked him what he made of Mrs. Welling's story.

The Vicksburg Ghost

"Well," he said thoughtfully, "when someone puts another person at the center of their life, they read about him, they think about him, I'm not surprised that he becomes real for that person."

Saturday night, at nine o'clock, Louise Welling is standing next to me in the Filling Station's parking lot—it is built on the site of John Vickers's flour mill—in a crowd that has just seen prizes awarded in the fifties dance concert and is waiting for the beginning of the second annual Elvis-look-alike contest. She is neatly dressed in a blue-and-white checked overblouse and dark pants. Her hair is fluffed up, and she is wearing pretty pink lipstick. She invited me to come to the contest, and told me that although many of the entrants in such affairs didn't come close to Elvis she was hoping that this one would draw the real Elvis Presley out from hiding. "If he came to me in the past, I believe he'll come again," she said. "I hope it will be before I die. If he comes, I'm going to grab him and hold on to him and ask him why he couldn't just be honest about needing to get away for a rest. Why couldn't he just tell the truth? Look at all the trouble he's caused those who love him."

Earlier in the day, I stopped in at Mar-Jo's for coffee. There were lots of extra visitors in the café. Ken Fowler had turned on the radio to WHEZ, a Kalamazoo station, which was broadcasting live from out on the street, acting as the festival's musical host. Rock music filled the café. Patrons were beating time on their knees, and the waitresses had begun to boogie up and down behind the counter. I asked one of them—a girl named Laurie, who was decked out fifties style with a white floaty scarf around her ponytail—what she made of Mrs. Welling's story. "I think it's kind of fun," she said. "I haven't met the lady, but, you know, maybe she's right. After all, if Elvis Presley never died he has to be someplace."

Mrs. Welling is subdued, as she stands next to me, but all attention—scanning the people, anticipatory. We are at the very back of the good-natured crowd, which has enjoyed the nostalgia, the slick cars, the dances, the poodle skirts, and the ponytails. She spots Kajkowski and says to me that he's not Elvis but "so far he's the only one here who even looks anything like him."

Skip Knowles is up on the stage, in charge of what has turned out to be a successful event. There have been record-breaking crowds. Six hundred and fifty cars were entered. He has had plenty of media coverage, and he seems to be having a very good time. He calls for the Elvis contest to begin. Ray Kajkowski's act is so good now that he has no competition—he is the only one to enter. I watch him play the crowd. He had told me, "When I first started, I really liked the attention, but now it's just fun to do the show, and, yeah, I do get caught up in it. I like the holding power I have over people. I know how it is to feel left out, so I play to everyone. But I like people in their mid-thirties or older best. I don't like to entertain for these kids in their twenties. The gals back off when I try to drape a scarf around them. I think that's an insult." Now he is dancing around the edge of the crowd, reaching out to kiss the women, who respond to him with delight and good humor, and then he launches into what Mrs. Welling tells me is "You're the Devil in Disguise." I look at her, and she seems near tears. Her shoulders slump. "I don't like to watch," she says softly, and walks away to gather her family together for the trip home.

On my own way home, on the morning after the festival, I made one final stop in Vicksburg, on the south side of town, at what is left of Fraser's Grove. For about forty years—up until the early 1920s—Fraser's Grove was one of this country's premier spiritualist centers. In 1883, Mrs. John Fraser, the wife of a well-to-do Vicksburg merchant, turned the twenty-acre woodland into a camp and gathering place for mediums, believers in mediums, and the curious. She had been inspired by a lecture on spiritualism given in a hall on Prairie Street by one Mrs. R. S. Lily, of Cassadaga, New York, a town in the spiritually fervent "burned-over" district of that state. In the years that followed, Mrs. Fraser became a national figure in séance circles, and another resident of Vicksburg, C. E. Dent, was elected president of something called the Mediums' Protection Union. A group calling itself the Vicksburg Spiritualists was formed shortly after Mrs. Lily's visit, and it met each Sunday. Its Ladies' Auxiliary held monthly chicken dinners (fifteen cents a plate, two for a quarter). On summer Sunday afternoons, people from around this country and abroad packed the campground at Fraser's Grove to talk of materialization and reincarnation and watch mediums go into trances to contact the dead. According to a 1909 issue of the *Vicksburg Commercial*, they debated subjects such as "Is the planet on which we live approaching final destruction, or is it becoming more permanent?" (A follow-up article reports that the Spiritualists opted for permanency.)

Trees still stand in much of Fraser's Grove, although some of them have been cut down to make room for a small housing development. The campground itself has been taken over by the Christian Tabernacle, which makes use of the old camp buildings. Tazzie, my German shepherd, was with me, and I parked at the edge of the grove to let her out for a run before we drove onto the interstate highway. We headed down a dim path, where events passing strange are said to have taken place. The grove produced no Elvis, no John Vickers, not even a phantom band concert or the apparition of Mr. Matz—no spirits at all. But Tazzie did scare up a rabbit, and the oaks were still there, and, untamed through 150 generations, so were the mosquitoes.

—*The New Yorker*, September 25, 1989.

Magic in Michigan

> Contiguity and succession
> are not sufficient to make us pronounce
> any two objects to be cause and effect.
>
> *A Treatise of Human Nature,*
> by David Hume, 1739–40

> Magic is a mind-expanding art form because
> it demonstrates that the boundaries of perception
> aren't necessarily the boundaries of reality.
>
> —Jay Scott Berry, California magician,
> at the Fifty-third Annual Magic
> Get-Together, Colon, Michigan, 1990

Colon, Michigan, is the magic capital of the world. You know that's true because the sign at the village limits says so:

> Welcome to
> Colon
> Magic Capital of
> the World
> 1832 Sesquicentennial 1982
> 1989 Class D
> State Baseball Champions

Colon is a village of eleven hundred people in southwestern Michigan, in flat prairie farmland. It is a village of two-story Victorian houses and smaller, modest ones, each nestled into an ample, well-tended yard, and there are gardens big enough to grow sweet corn two blocks from the intersection of the two main roads, State Street and Blackstone Avenue. The avenue is named for Colon's most famous past resident, Harry Blackstone, the magician: Blackstone the Great, né Henry Boughton, who came here in 1926, and for whom it was home, as much as any place could be for a traveling entertainer. It has more parks than anything else, and maple trees—big, overarching ones—really do grow on Maple Street. And there is not a place in the village where you can't hear the cooing of mourning doves on a summer day.

Colon is also the home of Abbott's Magic Manufacturing Company, founded by Percy Abbott, an Australian magician, in 1934, a few years after a Blackstone-Abbott magic company, started in 1927, failed, and Blackstone and Abbott had a falling out. With a million dollars in annual sales, made mostly through a three-pound mail-order catalogue, Abbott's is the biggest magic manufacturer in the world, according to the present owner, Greg Bordner. And for a time in late summer of each year since the 1930s there have been more magicians around Colon than anywhere else on earth. The occasion is the Magic Get-Together, co-sponsored by Abbott's and the Colon Lions. Magicians are a clubby lot and hold other meetings, but those are usually in big cities. According to Harry Blackstone Jr., a magician in his own right, the Get-Together is the meeting that all magicians look forward to, because of the rural setting and the air of sociability. A thousand magicians, perhaps a few more—some full-time performers, the rest part-timers and amateurs—come to Colon from around the world, and take over the village. For twenty years now, Blackstone himself has been coming back to his hometown (he went to school here for a year or two, and still owns a bit of land on what is called Blackstone Island, in Sturgeon Lake).

I grew up in Kalamazoo, about twenty-five miles to the northwest as the crow flies. I was the easiest kid in the world to fool; for a while, I cherished the notion that I had a head full of nickels, because my father could produce them so easily from my ears. In those days, I was a pushover for the public performances held in Colon in the evenings during the Get-Togethers. I can remember the aura of those evening shows, shows that I attended forty years ago and more: the dark stage, the bright colors, flowers, floating scarves, capes in black and red, and the daring, slightly wicked quality of dying vaudeville.

In August of 1990, I decided to pay a nostalgic visit to Colon to see what happens when a thousand magicians come to town. It was the Fifty-third Annual Magic Get-Together. Colon may be only twenty-five miles from Kalamazoo, but in order to get there I had to take out a map and look up which turns to take on state and county roads—a process lending substance to a conceit that Bordner and Blackstone like to maintain in interviews, that Colon is like Brigadoon, a hard-to-find place that disappears except during Get-Together time. I had known Harry Blackstone when we were both eighteen, and, seeing him recently in California, where he lives, I had asked him what growing up in Colon was like. "Shades of gray," he said, with a sigh. I knew what he meant. When I was growing up, I was under the impression that the sun vanished in October and was not seen again by humankind until May. The weather, blowing in from Lake Michigan, was bleak, cold, damp, and dark for much of the year.

On the first day of the Fifty-third Get-Together, however, Michigan was at its winsome best. It was a clear, sunny, high-pressure day, with blue skies and fluffy white clouds, and the corn was greening in the fields. In the village, I parked under a shady tree, because Tazzie, my dog, was with me. I walked over to the Abbott plant, a low cement-block building that would look like a radiator shop if it were not painted black with white skeletons dancing across its front. I passed a knot of people on the street. "How d'ya like this?" a man was asking. He shook his wrist, and bright red carnations bloomed from his fingertips. The members of his tiny audience nodded, smiled, and applauded. Inside the Abbott office I picked up my registration papers. I've never quite recovered from being fooled about those nickels, so I had paid the full professional fee, seventy-five dollars, to be allowed to attend not only the four public evening shows but also the dealers' showrooms and the workshops, lectures, and demonstrations. In my envelope of materials I found a plastic badge with my name on it and a cord attached, and I hung it around my neck to separate myself from the ordinary citizens of Colon. I peered inside the Abbott showroom. It was a dark room with creaky wooden floors, and its walls and ceiling were covered with magicians' posters, including one of the senior Blackstone, backlit to emphasize his aureole of white hair. The room was lined with glass cases that looked like old-fashioned candy-display counters but were filled with magic tricks, and in the center of the room an Abbott employee was putting on a magic show, mostly for children.

It was only ten in the morning, but magicians were already arriving in town, many of them in vans and oversized campers. Colon has no hotels, no motels, no inns. Some of the magicians camp out in the parks, but more rent rooms or houses from the residents. And many pairings of guest and landlord have stayed the same over the years, giving the Get-Together a mixed quality, of friendship and cash crop. Yard sales were springing up all over town, and I talked to the proprietor of one. She told me she was renting out a room to a magician couple who had been staying with her each August for the past ten years. "When my husband was alive, we used to rent out nearly every room in the house, and all our outbuildings, too," she said. "Haven't done that since he's been gone. But I've been thinking about all the campers that come these days. I guess I'll get me a batch of Porta Pottis next year and set them up and rent out the yard."

I walked up State Street, where all the merchants had stenciled signs in their windows that said "WELCOME MAGICIANS" alongside posters for a circus that was coming to nearby Three

Rivers later in the month, and out along the creek, past the dam that makes Palmer Lake. Beyond the dam I found the elementary school, where many of the activities were to take place. Just inside the school door, a uniformed security guard with a big smile was checking every plastic badge. The school gym and a couple of classrooms were filled to overflowing with magic wares and books, in odd contrast to the little desks and sober blackboards, and magicians were busily taking over rooms for workshops, demonstrations, and sales booths.

The magicians—most of whom were men—talked magic, passed out business cards, did magic, passed out business cards, bought magic, passed out business cards, demonstrated gimmicks, passed out business cards. "Gimmick" is a magicians' word that has entered everyday language. It refers to that device or object—that invisible wire, false fingertip, hidden lever—which makes the magic possible. Magicians use other words specially. Scarves are called "silks." "Disappear" and "vanish" are transitive verbs, as in "I vanished the ball" and "He disappeared the girl." "Tricks" are small pieces of magic—the silk that turns into a dove, the cut and knotted rope that mends itself. "Illusions" are big pieces of magic. Illusions are pricey; they can cost many thousands of dollars. Abbott's has made many of them over the years, some to order: Houdini's underwater torture box; Blackstone's buzz saw; David Copperfield's straitjacket; the device that disappeared an elephant, for which Abbott's furnished everything except the elephant. One whole wall of the gym was covered with a display of books: *How to Sell by Magic, Nite Club Illusions, Conjurors' Psychological Secrets*, and so on. Heaped on shelves along another wall were Abbott's offerings: Dove Through the Glass ($135), Elusive Bunny Box ($100), Mirror Tumbler ($5), Disecto ($90), and more objects in Day-Glo colors and shiny metal than I could take in.

Greg Bordner was standing near the Disecto, expansive, cheerful, busy. A handsome, athletic-looking man in his late thirties, he is a political-science graduate of Michigan State, and the son of Recil Bordner, a magician who was Percy Abbott's partner after Abbott broke with Blackstone. Versions of the cause of the split vary, but all agree that the two men, both strong-willed, clashed over the financial end of the business, and that Blackstone, the bigger of the two, came over to Abbott's house one night, beat him up, and stalked out. According to Bordner, Abbott called the police, and when they arrived Blackstone walked back in and greeted Abbott like an old friend—"Hi, Percy, it's been a while. Good to see you"—thus flummoxing the police. But, Bordner said, Blackstone stayed so angry that whenever he saw Abbott in the grocery store he would pull canned goods off the shelves and hurl them at him. He also tried to start a rival magic show and manufacturing company, and refused to perform for the Abbott Get-Together until 1961, when he was seventy-five. Though his skills were failing, he came from California to perform in a sentimental return; when he died, in 1965, his ashes were buried in the cemetery at Colon, where they lie under a stone that appears to be part flower, part flame. According to Bordner, more magicians—around a dozen—are buried in that cemetery, along with Blackstone, than anywhere else in the world.

I asked Bordner about the Disecto, and, confiding that he was only a fourth-rate magician, he got it out. It looks like a simple wooden stand with a big hole in the center and two little holes above and below. At the top is a sharp-looking knife that raises and lowers like a cleaver. The cleaver slices through carrots placed in the little holes. Bordner put his arm in the big hole and dropped the cleaver. His hand was just fine when he pulled it out. I gasped. I couldn't help myself. "How did you do that?"

Bordner grinned and said, "The trick's told when the trick's sold." He invited me on a tour of his empire, and we went back to the Abbott building. Behind the display room was a dark, Dickensian stockroom lined with drawers bearing labels like

"Rabbit Wringer" and "Flower Surprise." There orders were being filled for Sweden and Japan. We walked past a set of sinks where silks are dyed and the staff makes coffee, then went downstairs to a room where a woman was making magical silk flowers, stitching away and listening to country music on the radio. Outside again, Bordner put a big box of outgoing mail in the back seat of his Chevy ("The magic business isn't what it used to be. My father drove a Buick") and drove me over to the workshop, a nondescript blue-and-white metal building on the edge of town, next to the supermarket. Except for a preponderance of rabbits, it looked like an ordinary wood-and-metal shop. Tacked to the walls were snapshots of magician customers, including one of a Nigerian prince who once came, with an entourage, to the Get-Together. He placed a big order for illusions and had them shipped home. Bordner shook his head over the prince. "He billed himself as someone who had Powers," he said. Bordner doesn't approve of Powers; he doesn't like to associate magic with the paranormal. He made a face. "Hey, this is just a business. We manufacture tricks and illusions. I know the U.P.S. rates anywhere."

I very much wanted to buy a trick. Back at the elementary school, I asked around about the gimmick used for one of the world's oldest recorded magic tricks. In the Bible, in Exodus, we are told how Aaron so impressed Pharaoh with wonders and magic that Pharaoh let the people of Israel leave Egypt. Aaron showed Pharaoh a rod and changed it into a snake. Pharaoh called his magicians, and they also turned rods into snakes, but then Aaron caused his snake to eat up theirs. Bordner laughed when I asked him about it, and said he didn't have a trick like that. "Maybe one of the snake handlers could help you," he added. He looked around the gym-salesroom, but couldn't spot a snake handler at the moment. I moved on to another salesman, who was doing impossible things with what looked like a solid hoop of gleaming metal. It became a square, then a flat strip. "Oh, yeah, I know that trick," he said when I asked. "I got a wand that turns into a snake, but it's not a very good snake. I mean, it isn't a python or a cobra, or anything." I watched a New Age magician, Jay Scott Berry, who was dressed in black and had long, curly blond hair, produce a brilliant flash of light from his fingertip: "Uses no batteries." Then he transformed a rainbow-colored bar into gold.

"Ummm!" said a watching magician. "That's smooth."

"Ye-ahh," said Berry, changing from entertainer into salesman. "I worked a long time to get it so smooth." He was monopolizing the crowd, to the chagrin of a Japanese magician at the next booth, who smiled bravely as he plucked fire from the air and lit a fire in the palm of first one hand, then the other.

I settled for a trick called the Amazing Keybender ("Yes, you can bend keys and spoons"), which I bought from Abbott's for five dollars. It is a small, simple gimmick, and comes with three pages of directions which include the plea "All we ask is that you do not claim to have supernatural powers," and add archly, "One doing that is enough"—a prim reference to Uri Geller. I practiced the trick at home after the Get-Together, but no one applauded. "What do you have up your sleeve?" my stepson asked at my first and only family performance. That brings me to the basic secret of magic: The gimmicks are simple, but the tricks are hard, and require impressive skills and dexterity, and a keen sense of psychology for misdirecting the audience's attention. Every evening during the Get-Together, there was a public show, and during the day spontaneous shows never stopped: a thousand magicians showing off to a thousand magicians, receiving and giving applause and appreciation. They all knew the gimmicks. What they were applauding was the skill in their use, the quick hands and sly diversions, the breaking apart of cause and effect, forcing the senses to trick the brain.

Colon grows weirder and weirder. Everywhere magicians' vans are parked. Magicians nap in folding chairs. Magicians sell magic to one another. Magicians sit on curbs and throw their voices, so that trash cans begin to twitter like birds. At the counter of the M & M Grill, a stocky man in glasses hits his hand on the counter—*whack! whack!*—disappearing fuzzy balls every time. His companion nods appreciatively while he eats his cheeseburger. There would be applause from the booth behind them, except that one of the six magicians sitting there has the attention of the rest, asking them to pick a card from the fan of cards he holds in his hand. Jugglers fill the parks. I talk to a young juggler named Steve. "I do magic, too, but I prefer juggling," he says. "There are lots of jugglers around these days, because we had parents who were hippies who took up juggling; it was supposed to represent psychic balance, or something. We don't go in for that stuff, but we learned how to juggle from them. I'd like to get a job on a cruise ship or work in a theme park." Theme-park magicians, I learn at an afternoon workshop, are paid seven hundred dollars a week, and must do five shows a day, six days a week. Cruise ships pay from four hundred dollars a week to eighteen hundred; the occupational hazard is weight gain from good food.

Waiting for the evening performance, in the high-school auditorium, I take Tazzie out for a romp in the meadow behind the high school. She leaps from the car and begins running joyous circles around a young, sinister-looking man with punked hair dyed flat black, who is wearing black pants, slightly pegged, and a black shirt—the preferred dress of young magicians. With him is a savage-looking girl with purple makeup and a mane of black hair, who is wearing a black leather miniskirt. They squat down and scratch Tazzie on the belly, transforming themselves into teenagers. A clown is working out on a unicycle in the school-bus parking lot. Tazzie finds gopher holes to dig out, and I lie down in the clover, listening to the doves and the katydids. Overhead, swallows circle and swoop. How do they do that?

Later, inside the auditorium, women in sequined gowns

float in the air. One, to the amusement of the audience, refuses to be sawed in two and forces the magician to take her place inside the specially designed box. Ropes cut and heal themselves, and knots in them slide and disappear. Needles are stuck into balloons without popping them. Coins multiply. Cards are chosen and told. Silks appear from nowhere. A thousand magicians and six or seven hundred other spectators cheer and applaud under a banner celebrating the Class D Baseball Championship. (The team is called the Magi; its mascot is a white rabbit coming out of a hat.) The seats are hard and cramped, the sound system crackles and occasionally fails, but the performances are slick, smooth, very Big Time, paced for television. Last year's young-talent-contest winner has been invited back for a segment of an evening show: He presents a display of leggy girls, dancing, and other flourishes that looks as though he had MTV in mind. Other magicians use backdrops with their names in flashing lights and work to a rock-and-roll beat. Blackstone's segment is filmed for a TV show to be broadcast next spring. Magicians from Switzerland and Czechoslovakia are featured, calmer and wittier than the Americans, who are from a vaudeville tradition. A contingent of Indians comes into the auditorium, the women dressed in saris. I ask a gray-haired matron with an empty seat beside her if it is taken. "It's my husband," she says. "I vanished him."

One morning, I dropped in on the Vent-O-Rama. Ventriloquists call themselves "vents," and the Vent-O-Rama is a workshop for them. Ventriloquism has long been a part of the Get-Together. Edgar Bergen was born in nearby Dowagiac, and sometimes attended the Get-Together with Charlie McCarthy and Mortimer Snerd. "The *Pythonists* spake hollowe; as in the bottome of their bellies, whereby they are aptlie in Latine called *Ventriloqui*," Reginald Scot wrote, in 1584, in *The Discoverie of Witchcraft*, which is said to be the oldest surviving text on magic tricks. Not only does Scot discuss ventriloquism; he says that if one is not a "sluggard," a "niggard," or a "dizzard" one can learn to perform "Magicke," and proceeds to give detailed directions. While reading that four-hundred-year-old handbook, I was pleased to note that my father had been no dizzard: his nickel-from-the-ear trick is detailed by Scot under the heading "Of conveiance of monie." It is done by "palming," the word that magicians use to describe the way to hold a coin, card, or other small object hidden between the fingers or on the back or front of an apparently empty hand. You and I don't have muscles trained to do that, but they do. ("This is hard," said Doug Anderson at an afternoon workshop, as he demonstrated how to make a coin appear to jump straight up from his motionless hand. "It was painful to learn how to squeeze my palm muscles. Broke some blood vessels, too.")

Scot's may be the oldest handbook of magic tricks, but conjuring, or trickster magic, has an even older history. Aaron and the Egyptian magicians with their wand-snakes were not the only ones in antiquity to use gimmicks. (Aaron's most famous trick, making a rod bloom with flowers—by which he established the preeminence of the tribe of his brother, Moses, over the eleven other tribes of Israel—can be found in the Abbott mail-order catalogue: the Flower Wand, under "Flowers that bloom WITH a spring, tra la!," for four dollars.) What is thought to be the first self-moving vehicle was designed by Hero of Alexandria, an engineer who flourished around the beginning of the first millennium. It was used to make gods and goddesses move upon their altars, revolve, and produce milk and wine as if by miracle. Egypt, Greece, Persia, Rome, China, and India all had magicians, and some of their tricks—Cups and Balls and the Hindu Rope Trick, to name two—are part of modern magicians' stock-in-trade.

The morning I was at the Vent-O-Rama, the principal vent had a cold and a sore throat, so he asked the vents perched on tables and chairs around the classroom to volunteer. Almost before the words were out of his mouth, a young woman with red hair called out "Me . . . me . . . me!" and ran to the front of the room with two oversized stuffed toys. (Magicians are not shy. Boffo routine in the evening show: The MC says, "Tonight, we have the world's greatest magician with us. Will he stand up, please?" A thousand magicians leap to their feet, and everyone applauds.) I watched the red-haired ventriloquist as her stuffed toys carried on a lunatic conversation with her. Her mouth didn't move; her throat muscles didn't even quiver. How does she do it? When she returned to her seat, amid applause, to give way to the next performer, she regained our attention momentarily, and received some appreciative chuckles, by throwing her voice to the front of the room.

In the village, the good people of Colon were busily turning goods and services into dollar bills. In the general store, a troupe of young magicians was buying cotton balls. "Gonna have a new opening tonight," one of the women said. And, sure enough, that evening I saw those mundane cotton balls deftly appear and disappear between outstretched fingers.

I talked to Bob Kolb, the owner of Magic City Hardware, who is active in the Chamber of Commerce.

"You bet the Get-Together is good for the economy," he told me. "People rent out their houses. Some people even leave town for vacation and turn their houses over to the magicians. There are some real little houses here, and I've seen maybe ten magicians in a single one of them. Then, there are the fund-raisers. The Lutherans put on a big lunch, and the magicians like it, because they don't charge big-city prices. I sell the magicians nuts and bolts to fix their campers, and some big-ticket items, too. And then there are the merchants' sidewalk sales—How am I going to put this? We all have some stock . . . Now, it isn't bad merchandise, it's just that folks in Colon haven't bought it all year. These magicians come from all over—California, New York, Canada—and their tastes are, um, different, so we sell it to them. We unload a lot of stuff that way."

SUE HUBBELL

Craft sales began to appear among the yard sales. Soon it seemed that every third house had something for sale: wrought-iron trivets, odd dinner plates, hand-painted ceramics, little wooden benches, children's outgrown clothes, plastic purses. Curbside-parked cars had "FOR SALE" signs in their windows. Amish had driven in from farms in their buggies and were racing around town to snap up the best bargains before the magicians could get to them.

Having attended the Vent-O-Rama, I had missed the magicians' buffet lunch at the Lutheran church, a handsome, new, sprawling building on the outskirts of the village, right next to the Enchanted Glen Apartments. Instead, I went over to the Grange Hall, on Blackstone Avenue, for the Pig Roast. Inside the hall were long tables covered with oilcloth. A motherly-looking woman served me barbecue, baked beans, two home-made molasses cookies, and a cup of iced tea, for three dollars and ten cents. This was a fund-raiser. It was the first time the Grange had put one on, she told me. "Some years back, the Lutherans used to serve dinner every evening during the Get-Together. That's the way they raised the money to build the church." The church that magic built.

After a whole day of magic and a three-hour evening performance, the magicians, their attention sharpened, tingling from their own applause, are high on magic, and reluctant to end the day. When the tourists (some three thousand of them paid to see the public shows during the week) have driven away, the magicians return to the Abbott showrooms for auctions of magic equipment, or they go over to the American Legion Hall, on State Street. I drop in at the hall about midnight. It is stuffed with magicians drinking beer, eating fries, laughing, and showing off. A T-shirted magician moves from table to table performing what I have learned to call closeup magic. He is applauded as he moves a safety pin from one corner of a square of red flannel to another without unpinning it.

I find a place at a table with a group recalling their days on the road.

"D'you remember the night at the motel when we took the ice out of the ice machine and stuffed Ed inside?"

"Yeah. Remember the look on the face of that little old lady when she opened the machine door and he handed her a bucketful of ice?"

"Hey, how about the time in the hotel in Kentucky when we took the furniture out of our room and put it in the elevator, so that it looked like a sitting room, and rode up and down in it reading newspapers and watching the faces of the guests when the door opened for them?"

"Where was it that we filled up the motel swimming pool with Knox gelatin?"

The laughter grows louder; the air is thick with cigarette smoke, magic, and fellowship. At 2:00 A.M., the American Legion will close its doors and force the magicians out onto the street. They tell me about the young Doug Henning, seventeen or eighteen years old and unknown, who wandered around town one night after the hall closed until he found a twenty-four-hour Laundromat, the only public building with lights still on, and did magic tricks for his friends until dawn. His later fame has assured that it will forever be known as the Doug Henning Laundromat. Some years ago, they tell me, there was still enough of a crowd on the street at 3:00 A.M. to respond to the request "Pick a card any card." And, just as someone in the small audience did so, and looked expectantly at the magician to identify it, a heavyset man lurched toward the spectators, stumbled on the curb, and fell. Several men reached out to help him up, and as he regained his feet his pants slithered to his ankles to reveal baggy undershorts with the card printed large upon them.

As the week progresses, there is far more magic on and off the streets of Colon than I can take in. The wild mourning doves are reproduced onstage in the white doves that flutter

SUE HUBBELL

from silks, flowers, hats. A man balances 135 cigar boxes on the end of his chin. Wands shoot confetti, and it settles in our hair. White bunnies are everywhere. Street musicians play. A lank-haired man in a polo shirt walks down State Street tossing up balls, and they vanish high in the air. "Hey, I like it. I like it," his companion says, slapping his knee. One evening, Harry Blackstone re-creates one of his father's most famous performances. Taking a lighted lamp, with a stagily long electric cord, offered to him by his assistant—his wife, Gay—he removes the bulb. On the dim stage, the bulb remains shining eerily in his hand. He then apparently releases it, and it hovers dreamily about him in the darkness. He sends it, glowing, down to the front rows of the audience, where he invites spectators to touch it and reassure themselves of its substance. When a doubter farther back in the auditorium calls out, Blackstone says, with a smile, that he will send the bulb out to him. Luminous, spectral, the bulb—a mere light bulb, made lovely and strange by behaving in a way that no mundane light bulb should—drifts gently out over the audience and then soars back to its master, who returns it to the lamp. And I understand that I have never really seen a light bulb before.

One hot afternoon, I take Tazzie down past the dam to a little creek that runs from it. Tazzie wades into the water while I sit on the grassy bank and watch a pair of tiger swallowtails circle in the air above me. A pretty little girl with curly chestnut-colored hair comes walking down the bank from one of the houses nearby. "What's your dog's name?" she asks.

"Tazzie. What's your name?"

"Jessie. You know what?"

"What?"

"I'm four years old. Can Tazzie do tricks?"

"Not really. She likes to take rocks out of the water, though." I point to Tazzie, hard at work loosening from the bottom of the stream a rock much too big for her.

"Well, I can do a trick."

"You can?"

"Yes. I can put my head under water." And—hey, presto!—she does. Carefully pinching her nostrils, Jessie puts at least half an inch of her face into the water and very quickly pulls it out. She looks at me, expectant, proud. I applaud.

—*The New Yorker*, November 12, 1990.

Polly Pry

The scene was straight out of *Brenda Starr*, the comic strip. It was January 13, 1900, and the place was the so-called Bucket of Blood, the joint office of Harry Heye Tammen and Frederick Gilmer Bonfils, owners and editors of the *Denver Post*, who were on the floor bleeding from gunshot wounds. The gun had been fired by William W. (Plug Hat) Anderson, a Denver attorney. Tammen had just called him a son of a bitch and a robber, possibly even a cheapskate and a liar. Plug Hat was evening the score. The fourth person in the office was the tall, strikingly beautiful, fearless woman reporter (a blonde, alas, not a redhead) known as Polly Pry. Her series of stories on Colorado prisons and Alfred Packer, a man she believed to have been wrongly jailed for murder (he was also accused of cannibalism), was the cause of the dustup. After her bosses fell to the floor, she called to the newsroom for help. No one came, so she covered Tammen with the swirls of her dress to protect him—never mind that he was bleeding all over its hem—and grabbed the barrel of Anderson's gun. The journalists cowering outside the office door heard him threaten to kill her.

"Go ahead," she said. "And then hang." Polly Pry *did* love the dramatic moment.

But Anderson didn't kill her. She lived to outrage others with her writing and to look down the barrel of a gun again. In the end Plug Hat Anderson walked out of the office and turned himself in to the police, saying, "Arrest me. . . . I just killed two snakes." He hadn't killed either editor. Tammen, in fact, was lively enough to object when the doctor dressing his wounds began to take off his shirt. "Don't cut it, damn it," he grouched. "It's silk."

At the time of those difficulties in her editors' office, Polly Pry was forty-three, a locally famous writer of flamboyant, investigative newspaper pieces. She was gorgeous, a witty conversationalist, a star of Denver society, and she was full of beans. Twenty-five years later she would be named in *Liberty* magazine as one of the most interesting women in America.

She was born Leonel Ross Campbell in 1857 on a plantation in Mississippi. Her parents, Mary and James Campbell, sent her to boarding school in St. Louis, a period of her life that she later characterized as "the silver years of laughter." Silver or no, in 1872, when Leonel was fifteen, she donned her first long dress, a black velvet, and climbed over the school walls to elope with George Anthony. Many of the details of her life are tantalizingly vague, and how she met Anthony and who he was are some of them. He was said to be "of the notable Kansas Anthonys." He took his bride to Mexico where he had a commission from a group of Boston bankers to build the Mexican Central Railroad.

Sometime later she described her honeymoon in this characteristically florid fashion:

> I was a bride of only a few days' standing when I arrived in El Paso . . . with my better half, who was an official of the Mexican Central Construction company and had been compelled to hurry back to his post. Rooms had been prepared for us at the hotel, but when that first hot night I was discovered, about midnight, seated, à la Turk, on the top of a center table, with the tears streaming down my cheeks and the whole place reeking with the suffocating odor of pennyroyal, of which I had used a quart, I was simply bundled into a carriage and taken over to the private car, where I lived for many long days before I could be persuaded to return to my rooms where the fleas had so completely routed me.

Leonel left Anthony after only a couple of years. What happened next remains a mystery. The story handed down—though not completely substantiated by facts and dates—is that following the breakup of her marriage she went to New York City, where she asked a family friend, John Cockerill, editor of the *New York World*, for a job. "Give you a job!" he is reported to have said, "I ought to spank you and send you back to your husband." But the youngster turned in such a good story about a fire that she established herself on the *World* and went on to become its Latin American correspondent, covering Panama when there was just talk of a canal. On the side she wrote pulp fiction, which was published by Street & Smith.

Years later she would write about what she had learned in those days:

> The first time I ever interviewed a man they sent me out to ask a Bishop of the Greek church whether he had set fire to the church building and the Boys' Home, and nearly burned some score or so of little boys to death in their beds, stolen the church books and embezzled moneys put in his charge. I sat in the church parlor and watched the Bishop come down the stairs to be interviewed. He was a man six feet and a half tall. He wore a trailing black robe, a tall pointed cap, and his hair fell in shaggy curls to his waist. Also his beard. I didn't run. I stayed and asked him all about it. I didn't know any better.

An editor once advised her: "Tell the story as it is, and remember, on this paper you are not to have any veneration for anybody or anything except God—and—well you might temper that a little."

Her parents had moved to Denver, and visiting them there in 1898 she met Bonfils who, with Tammen, had taken over the *Denver Post* in 1895. He offered her a job to write as she pleased. She took it and, probably at the suggestion of Bonfils and Tammen, began calling herself Polly Pry. That pseudonym echoed the pen name Nellie Bly, which was used by Elizabeth Cochrane Seaman and made famous by her around-the-world story and exposé journalism. Polly Pry was equally footloose. She covered Colorado politics and reported firsthand from Moscow and points in between.

The *Post* was building its circulation on populism and opposition to authority, all authority. So after Polly Pry met Alferd (sometimes spelled Alfred) Packer in the course of writing a prison series, she and Bonfils and Tammen decided to crusade for his release. Packer had been convicted in 1883 of killing five prospectors he was guiding over a high plateau in Hinsdale County in southwestern Colorado. Packer eventually admitted to eating the men to save himself from starvation and to killing one in self-defense. The judge in the case, a fiery Democrat, delivered what was said to be an eloquent hanging speech, full of classical allusions, but what has come down from it is the report of a spectator who, liberally juiced on "Taos Lightning," staggered into a nearby saloon where he called out: "Well boys . . . Packer's to hang. The Judge, God bless him! says, says he: 'Stand up, yah man-eatin' son of a bitch. . . . They was siven Dimmycrats in Hinsdale County, but you, yah voracious . . . son of a bitch, yah et five of thim! I sintince ye t'be hanged by th' neck ontil y're dead, dead, dead; as a warnin' ag'in reducin' the Dimmycratic popalashun of th' State.'"

Packer's sentence was reduced to forty years in prison on a technicality, but even that was unjust according to Polly Pry. (In 1989, a forensic scientist exhumed the five bodies and concluded that Packer had indeed killed and eaten his victims.) She wrote a series of columns citing cases of those convicted of murder and rape who were given lesser sentences, and finished each one with "Why not Packer?" In early January 1900, she was approached by a hustler who was planning to open a tobacco stand and hoped to hire a free-and-famous Packer to

SUE HUBBELL

run it. It was he who suggested Plug Hat Anderson as an attorney to bring about Packer's release. There is no report of what the hustler paid Anderson, but Polly Pry offered him $1,000 of *Post* money to take the job. He apparently accepted that and, misrepresenting matters, extracted from Packer all the money that the cannibal had earned in prison making hair ropes and bridles. And it was a quarrel over those latter funds that brought Anderson to the Bucket of Blood. The office of Bonfils and Tammen did not acquire that gruesome designation after they were shot there; it was already known as such because its walls were painted red—a color that matched the *Post*'s combative editorial style.

Packer was freed in 1901, about the time Anderson was being tried. During the proceedings, Tammen was charged with trying to buy the jury, and the litigation went on and on. At some point, a distinguished Denver attorney named Harry J. O'Bryan entered the case as amicus curiae, an adviser to the court, in which capacity he probably met Polly Pry. O'Bryan was a conspicuously fashionable man about town, dapper in a gray suit and white vest, with trouser creases so sharp that they received notice in the local press. Remember O'Bryan. He's plot material.

Eventually Polly Pry's sense of property and aspirations of class affected her reporting and forced her into a clash with her employers as she began covering what might be called the Colorado Mining Wars—conflicts that have made Telluride and Cripple Creek infamous in the history of the American labor movement. The fortunes of great families, the Guggenheims, Rockefellers and Goulds, were built in part on the valuable ores that lay under the Colorado soil. In their extraction, greed and a sense of frontier freedom bred violence from both labor and capital. William (Big Bill) Haywood, who later helped found the Industrial Workers of the World and backed Eugene Debs as Socialist candidate for President, was radicalized in the Colorado Mining Wars. Mary Harris

Jones, the radical labor organizer known as Mother Jones, spoke and inspired miners there as she did in Pennsylvania and West Virginia. In a disagreement over a story she had written about the president of the union, Polly Pry parted company with Bonfils and Tammen, who were more sympathetic to labor than she was.

Polly Pry believed in turning disappointment into triumph, so she started her own magazine, a weekly focusing on local and regional affairs. She named it after herself, *Polly Pry*. The first issue, dated September 5, 1903, shows her on the cover, an elegant woman with an hourglass figure in swirls of lace and a knock-'em-dead hat. It may have been one of the liveliest, most libelous, most outrageous magazines ever published. In it she would express her assorted social views: enduring opposition to anarchism, socialism and the municipal ownership of utilities, as well as to trade unionism. As she saw it, "The fruits of production belong to whoever has the brains and the money to put in force a plan or a business that will produce." She stood for feminism and women's suffrage.

She must have seen herself when she wrote about "a brainy woman who, when she found she had no vocation for wifehood and less call to fill the office of mother, tarried not . . . but went her happy unmolested way." She wrapped her social views in lively features and gossip. The latter is delicious to read when you have heard of its subjects:

> They say Mr. William Jennings Bryan offered the fair Ruthy $25,000 and a three-years' trip to Europe if she would give up marrying the picture man. . . . The picture man is over twice Ruth Bryan's age, and they do whisper that he is rather a naughty thing in the Paris studio sort of way.

It is delicious to read even when you've *never* heard of them:

> Everyone who knows Miss Lou Martin . . . will be glad to learn of her engagement to a Mr. W. K. Bender of New York. All the

romantic tales which made Miss Martin out the heroine of a broken heart idyll will have finis written at the end of them at last. Miss Martin is the fortunate girl who had the narrow escape from one Edward Ellis Hamlin, who eloped with Mabelle Hite, the Telephone Girl.

Two winters ago Mrs. Brown hankered for culture, and sat the lights out at Carnegie Institute learning things. After that Japan. She has come back with "Darling of the Gods" writ large on her face and a Yoe San arrangement of coiffure that gives one the backache.

Senator Carey has cut son Bobby off [without] a shilling and all because he married Julia Freeman. . . . Now Bobby will have to work to support wifey and work overtime, too, for Julia has tastes not at all compatible with the love in a cottage idea. It's ten to one that she married Bobby Carey on a dare. Two seasons ago she set the staid society of Santa Fe on the ears when she went to visit her cousin, Governor Otero's wife. . . . She has a pair of brown orbs which simply make a man get busy. She introduced "Bridge" into Santa Fe and collected her earnings, which were plenty, and was paid in silk hosiery. So much toward the trousseau.

As for Judge Wilson—well, really. . . . He has a front piece like a fried doughnut and the magnetism of a deep sea clam.

Remember Harry J. O'Bryan? The dapper attorney? In the September 23, 1903, issue we read:

Mr. and Mrs. Harry J. O'Bryan have gone to New Mexico to settle some matters. . . . Mrs. O'Bryan was at one time the life of the Overland golf links but her numerous visits to the East and poor health lately have caused her friends to be robbed of her delightful company.

Polly Pry covered sports:

> Denver University is going to give the Athletic Club a chase for their laurels this winter on the gridiron. . . . The biggest man on the eleven is Bert Martin, carrying 220 pounds as gracefully and as easily as a danseuse wears her duff-duffs.

Polly Pry covered culture:

> His interpretation of the Brahms Mainacht was ludicrous and his German unmentionable, but the audience was goodnatured . . . and gave him a rousing encore.

Inside this gossipy gift wrap, *Polly Pry* covered the Colorado Mining Wars. The editor was a staunch supporter of Governor James H. Peabody, who ordered in a thousand troops paid for by the Mine Owners' Association. The troops were led by an ex– Rough Rider named Sherman Bell. One reply to the charge that his treatment of strikers might be unconstitutional was, "To hell with the Constitution! We aren't going by the Constitution. We are following the orders of Governor Peabody." Labor lost the Colorado Mining Wars and, although it is true that the militia and interunion rivalry had a lot to do with it, Polly Pry, with her hard-hitting stories and her version of the "truth," certainly aided and abetted the forces of property and capital.

No history of those strikes can be quite understood without reading *Polly Pry* and understanding how forcefully its editor tried to influence her contemporaries. She was relentless in her attacks on strike leaders, but of particular note is her virulent opposition to Mother Jones. In a sense, the two women complemented each other. They were of the same time (Polly Pry, 1857–1938; Mother Jones, 1830–1930). Mother Jones was a college graduate who hid her culture. Polly Pry was a secondary-school dropout who was more like those she derisively labeled "culturines" than she would care to admit. Mother Jones was pro-labor. Polly Pry was anti-labor. Both were independent at a time when women were often dependent. Both were outspoken women who bent the truth to accomplish their own ends. It was predictable that they should clash. When Mother Jones came to Colorado to help organize the miners, Polly Pry fulminated: "Were there not enough homegrown noxious fungi to poison the Colorado miner's . . . broth that foreign toadstools must be imported?"

Polly Pry had social connections to the rich and powerful and in some undisclosed way the Pinkerton files were opened to her to provide some grist for what became her most scandalous story. Today agents of the Pinkerton company, whose private espionage service in the Civil War led to the establishment of the Secret Service, are security guards at banks, office buildings, and other commercial establishments. At the turn of the twentieth-century, they were routinely in the employ of corporations against strikers. It was from Pinkerton files Polly Pry took the information that, in her issues of January 2 and 9, 1904, she said proved that Mother Jones had been both a prostitute and a "procuress," charges that she supported with dates, addresses, and names. The story was widely reprinted and stuck to Mother Jones throughout her life and beyond it.

The question occurs as to whether Polly Pry was being supported by the Mine Owners' Association or by other wealthy, powerful patrons. But nothing in her papers, which are in the Western History Collection at the Denver Public Library, indicates that this might be the case. The papers are loosely organized, awaiting the future thesis or novel writer (whose book is to be liberally supplemented, please, by reprints of the woman's own writing—*The Best of* Polly Pry). Photographs and billets-doux (one from a woman) tumble out of file boxes together with subscription lists, newspaper clippings, scrapbooks filled with her stories. All the odd papers of a lifetime. There is a stock certificate for a Cripple Creek gold mine, but it must have been worthless or it would have been passed on to heirs. Most telling are her letters seeking funds for her many enterprises, the sad documentation of her scrappy, lifelong hustle for cash. There is even one odd passage in 1924 when she tried to act as a go-between in the sale of an extensive collection of European paintings to a buyer who would donate them to the Denver Art Museum, out of which she hoped to realize twenty-two thousand dollars. ("I am paralyzed for lack of money," she wrote, and "Nothing is impossible today except happiness and that has always been a myth.") The sale never came off, according to the museum.

Her publication is full of promotional gimmickry, including what must be one of the most unusual come-ons in the history of magazine publishing. During April of 1904 *Polly Pry* offered to "send the boy who gets the greatest number of subscribers . . . to Harvard, Yale or Princeton . . . the girl . . . to Vassar, Smith or Wellesley." This included not only four years of tuition and transportation but a ten-dollar-a-week stipend to boot. There is no mention of winners, however, in any of the subsequent issues.

This followed a January 1904 assassination attempt on Polly Pry that some of her contemporaries suggested might have been staged in order to increase circulation of the magazine. She denied it and said that labor was trying to silence her. Whatever the motive, on the evening of January 10, she warily opened the door to her house on West Colfax Avenue (later razed and covered by the grass of Denver's Civic Center), standing behind the door as she did so, thereby avoiding the two .45-caliber bullets that whizzed past her and slammed into the wall and sofa. Her assailant, whom she did not see clearly, was never apprehended.

"What is my crime?" she asked in her magazine. From that point on, her prose began to take on a shriller, grandiose quality:

> Colorado is the chosen battleground for the greatest industrial war ever waged in America. I'm going to tell the truth about

this war. . . . If I die in the telling of it—someone else will take my place.

So said Polly Pry in her January 23, 1904, issue, and the publication kept up the drumbeat until August 26, 1905, when the cover carried the announcement that there had been a change of management and that the magazine would, in the future, be issued as a "high class" weekly under the title of *The Saturday Sun*. "It will be sparkling, but [it will] not sting. It will be honest and fair, not vindictive." Notwithstanding all of that, it is uncertain whether a single issue of *The Saturday Sun* ever saw the light of day.

Leonel Ross Anthony dropped from view for the next few years. Among her papers, however, there is a 1907 letter addressed to her in Staten Island from a New York actress who was turning down a play that Polly Pry had written. There are a number of plays among her papers, all of them preachy, all of them bad. In 1910 she married lawyer Harry J. O'Bryan, he of the sharp trouser creases. There is no record of what the marriage meant to either of them, but in 1914, at the age of fifty-seven, she was sent by the *Denver Times* to look for Pancho Villa, the Mexican revolutionary, and on the eve of her departure she had word of O'Bryan's death in Oregon. That didn't stop her from entering into her new job as war correspondent with enthusiasm. Her reports contain some of her liveliest writing. When she eventually caught up with Villa, she wrote her impressions:

> Pancho Villa, illiterate, ignorant, arrogant, bloodthirsty and cruel, but a man who laughs at death, who knows no fear, who meets you with narrowed eyes of hatred and wide smile of friendliness and who must be reckoned with before peace is even thought of. . . .

For a time after that, she wrote music and drama reviews for the *Times*. She tried to establish a theater, the Lakeside Casino, in Denver. Her usual problem, lack of funds, seems to have brought this enterprise to a close. (Among her papers there is a fussy letter from a printer demanding prepayment of his bill, perhaps because of a bad credit record.) In 1918 she wrangled a job with the American Red Cross and, from the London office, served in Albania and Greece writing press releases. In 1921, when she was sixty-four, she applied for a job in Washington, D.C., with the U.S. Joint Committee on Reorganization. She did not get the job but stayed in Denver, where she spent her remaining years doing public relations work for the Red Cross and a livestock show, and hustling those European paintings. It was a long way from her glory days as Polly Pry, but as she had once written, "The woman who knows enough to earn a decent income for herself is very apt to have to earn it."

Leonel Ross Anthony O'Bryan died on July 16, 1938, at the age of eighty-one in a Denver hospital. Her final words, as she raised herself from the bed, were, "I must be up and . . ." Brenda Starr herself couldn't have said it any better.

—*Smithsonian Magazine,* January 1991.

Earthquake Fever

On Monday, December 3, 1990, there was no earthquake along the New Madrid Seismic Zone, an ancient weakness in the earth's crust which runs crookedly for 125 miles from northern Arkansas to southern Illinois, alongside the Mississippi River. Iben Browning, an inventor, business consultant, and self-taught climatologist, had said there was a one-in-two chance that on or about that date a major earthquake would occur in the area. No, said government and academic earthquake specialists: The chance was more like one in sixty thousand.

It was a non-earthquake about which no one was wrong. Browning could take comfort in the *other* one out of two, and the academics and government people could take pride in their 59,999. Of course, it's not quite true that no one was wrong. The St. Louis psychic who said that the earthquake would happen at the end of November, perhaps on Thanksgiving weekend, was wrong. A man named Jacques called in to a Rolla, Missouri, radio show and said that the earthquake would take place on Tuesday, December 4, because earthquakes were triggered by underground nuclear explosions, and his research showed that those explosions always happened on Tuesdays. He was wrong. So was Larry Evans, who wrote a letter to the editor of the *New Madrid Weekly Record* in which he said that while praying he had received the information that the earthquake would be at ten in the morning on December 2. And so were the coffee drinkers in cafés who scribbled with ballpoint pen "1234567890" on paper napkins, circling the first two numerals for December, the next for the day. Four-fifty-six was the time of the magnitude 7.8 earthquake. And then they would proudly circle the last digits, the nine and the zero, for the year.

I have a farm about a hundred miles from New Madrid, Missouri, which is in the middle of the New Madrid Seismic Zone. The zone is a seismically active area, and geologists expect that it will someday produce an earthquake of the dimensions of a series of earthquakes that took place there in the winter of 1811–12, which were among the most severe to have occurred on the North American continent in historical times and also among the most severe ever to have taken place on the planet. The zone continually acts up. There are several hundred earthquakes along it every year, most of them too small to be felt. But one afternoon some years ago I was standing in my living room and felt the floor jiggle beneath my feet. It made me distinctly queasy to have my most basic assumptions about the solidity and rightness of the world denied by the earth's turning to Jell-O. I learned later from the radio news that the earthquake was in the New Madrid Seismic Zone.

I first heard of Browning's prediction early in December of 1989. Browning had spoken at the Missouri Governor's Conference on Agriculture, held at Osage Beach from the tenth to the twelfth of December. News reports identified Browning as a climatologist from Albuquerque who claimed to have predicted the destructive San Francisco Bay area earthquake of the previous October. His new prediction was unsettling. I hadn't known that it was possible to predict earthquakes. I wondered if he could really do it. And, if so, how?

I telephoned Browning in early March, 1990. He sounded wary and defensive, and made it clear that he didn't like to talk to journalists, but he did explain to me that his "projection" (he bristled at the word "prediction") was based on tidal forces. On December 3, 1990, he told me, the sun and the moon would be in line with the earth, creating high tidal forces. "And high tidal forces trigger the eruption of volcanoes—I'm a vulcanologist, really; this earthquake business is just a sideline," he said. I told him I found that interesting and would like to see some documentation of the projection of the San Francisco earthquake before it happened. "If you want to know what the scientific community thinks of my work, you can ask David Stewart, the director of the Center for Earthquake Studies, in Cape Girardeau, Missouri," he responded, sidestepping my request. "He'll tell you how much my work is worth. And those other details . . . Well, I'm retired now, and my daughter, Evelyn, is editing my newsletter. You can talk to her." He handed the telephone over to Evelyn Browning Garriss. I needed some proof of her father's "projections," I told her. "Oh, you want something on his methods?" she asked. Well, yes, that, too. But it would be really important to show that he had mentioned the earthquakes *before* they happened, I explained. She took down my name and address and promised to send me a packet of information. It never came.

Soon a tape of Browning's speech which I had requested from the Missouri Agriculture Department arrived, and I was able to hear him say, "December 3, 1990, at thirty degrees north latitude, there will be the highest tidal force in twenty-seven years. . . . We are back in the same triggering-force configuration as of December third next year as we were when the great New Madrid earthquake went off. Will it go off? . . . If you pull a trigger on a gun it goes off only if it is loaded. . . . Certainly the opportunity exists for it to be loaded, because a hundred thousand cubic miles of dirt or so has washed off of this part of the world down to and is now constituting the Mississippi River delta. So there has been an enormous load taken off this part of the plate." He also entertained the audience with jokes about journalists and communism. And he spoke about the theory of the greenhouse effect: "Fundamentally, it's garbage." Browning believed instead that the planet is cooling. "By 2010 . . . the Mississippi will normally freeze over down to about Natchez." He tied historic cold periods to political unrest and predicted that the near future would be a time of famine, revolution, and war.

In May, I drove over to New Madrid to look around. New Madrid was laid out in 1789 by George Morgan, a Revolutionary War patriot who had grown disillusioned about the new republic's chances of survival. Working with Spanish settlers in New Orleans, Morgan drew up plans for a large city, a strategic metropolis, on the Mississippi River, which would be controlled by the King of Spain. The spot he chose was L'Anse à la Graisse (Cove of Grease, so called because of the fat bear and buffalo that could be hunted nearby), and, hoping to ingratiate himself with the king, he named the town New Madrid. There were difficulties. Morgan lost interest and returned to the East. The town survived, but not on the scale that Morgan had dreamed of. The Spanish built a fort, levied duties on river traffic, and by century's end the sandy bluffs along the river were home to six hundred people inured to fighting ague and floods.

That original New Madrid crumbled into the river during the 1811–12 earthquakes. The New Madrid of today is slightly to the north of the original site, in the part of Missouri called the bootheel, hunkered down behind the earthen levee that protects it from the Mississippi. Yellow flowers—hop clover, cinquefoil, and dandelions—were blooming on the levee the day I was there. They looked cheerful, but the river—muddy, powerful, turbulent, ominous—did not. Next to the levee is a historical museum, housed in a nineteenth-century brick building that was originally the First and Last Chance Saloon, a stopping point for thirsty riverboat men. In it are displays relating to Indians, to New Madrid's role in the Civil War, and to the great earthquakes. The museum does a brisk business in earthquake T-shirts ("It's Our Fault" and "Visit Historic New Madrid While It's Still There"). A docent told me that she was aware of Browning's prediction but didn't credit it. "We're used to earthquakes," she said. "We have five a week. Had one yesterday. I've lived here all my life, and I never even heard about those big ones until I was all the way grown up." The center of town looks rundown and shabby, and many of the stores along Main Street are empty. But beyond them are the well-tended houses of the town's thirty-three hundred residents, many of whom work for the Noranda Aluminum smelting plant or for Associated Electric Cooperative, a power-generating station, or for the A. C. Riley Cotton Company. This is primarily an agricultural town. Soybeans, corn, and cotton grow in the flat, alluvial bottomland that surrounds it.

That day, high winds blowing over the plowed fields turned the air yellow with dust. I drove on down to the Mississippi River crossing—past Portageville, which was originally Shin Bone and is now the home of the annual National Soybean Festival. I wanted to see Reelfoot, the lake created by the earthquake, on the Tennessee side of the river, catty-corner from New Madrid. Part of Reelfoot is now a federal wildlife refuge, and the day I was there it was specializing in great blue herons. The lake is fourteen miles long, five miles wide, and eighteen feet deep, with something about it that reminded me of an Army Corps of Engineers project. I had read an account of a hunter and trapper who was present at the time of the earthquake. He told of great fissures appearing and the ground sinking. I tried to imagine the sound and shudder made by the land as it gave way.

One Indian legend has it that Kalopin, or Reelfoot, was a Chickasaw chief with a clubfoot, who fell in love with a Choctaw princess named Laughing Eyes. Her father refused to let him have her, so he abducted her, despite a dream he'd had in which he was warned that the earth would rock in anger and the waters would swallow his village. During the marriage ceremony, the earth began to roll in rhythm with the tomtoms, and the Father of Waters roared over the village. Under Reelfoot Lake, the legend says, lie the bodies of Reelfoot, his people, and Laughing Eyes.

Earthquake Fever

But the Creek said otherwise. They said that when Tecumseh, the great Shawnee chief, was trying to form a federation of tribes to stop the white men from taking their lands he grew weary of their indecision and warned, in the autumn of 1811, that he would leave them and in thirty days' time would stamp his foot. He did so, and the earth rumbled and split, and then they knew that the Great Spirit had sent him.

The logarithmic Richter scale for measuring the magnitude of earthquakes was not developed until 1935, and so could not be used to measure the 1811–12 earthquakes. But the modified Mercalli scale, which measures the severity of an earthquake in a particular area—how many houses were destroyed, how much sand and mud were ejected from the earth—can be applied, using contemporary descriptions. It is expressed in Roman numerals, I through XII. The XII stands for total destruction and corresponds roughly to the high eights on the Richter scale. The great 1906 San Francisco earthquake, for example, was a VII. Starting on December 16, 1811, and extending through the following February, the New Madrid Seismic Zone experienced at least three earthquakes that are rated XI and one that is rated XII. In between, and for months afterward, there were almost constant smaller shocks. Samuel Mitchill, who was a professor of natural history and a senator from New York, collected eyewitness reports of the earthquakes, and told of a Dr. Robertson, living in nearby Ste. Genevieve, who recorded more than five hundred tremors and "then ceased to note them any more, because he became weary of the task."

In the western third of North America, the convoluted folds of the earth's surface and its fractured geologic structure tend to absorb the seismic energy of an earthquake. Even an 8.5 earthquake in Los Angeles, say, would fade by the time it reached San Francisco, some four hundred miles away. But in the eastern two-thirds of the continent the same energy travels more easily. Below the New Madrid Seismic Zone lies a deeply buried rift in the continental plate, a weakened spot where the earth's crust once started to pull apart. The rift is under stress from tectonic forces. A trough above the rift has, over millions of years, filled with gravel and sand. During those 1811–12 earthquakes, water mingled with the soft substrate and turned it into something resembling quicksand—a process called liquefaction. Seismic waves moved up through the mixture and through the solid slab that is the surface of the eastern two-thirds of the continent. The earthquake, first felt in New Madrid on December 16, disturbed Washington, D.C., more than eight hundred miles away. According to Mitchill, "a gentleman standing in his chamber at his desk and writing, in the third story of a brick house, upon the Capitol Hill, suddenly perceived his body to be in motion, vibrating backward and forward, and producing dizziness." Tremors were also reported as far away as Boston and northern Canada. Judge James Witherell, of Michigan, told Mitchill that Indians living on an island in a lake outside Detroit saw the waters of the lake appear to "tremble and boil like a great pot over a hot fire; and immediately a vast number of large tortoises rose to the surface, and swam rapidly to the shore, where they were taken for food." All told, tremors were felt over an area of a million square miles—half of the United States.

There is no lack of firsthand reports from New Madrid itself, for nearly everyone there survived. Understandably, the accounts vary according to the place where the witness was shortly after 2:00 A.M. on December 16, when the first shock hit, but all the witnesses speak of feeling terror and nausea as the earth heaved, and describe the earth being torn apart, the elements mingling in a frightening way, and hearing sharp noises and rumbling ones as the earth spewed sulfurous gases, sand, and coal. The air, thick and malodorous, was colored by flames from small cabins collapsing into hearth fires. The Mississippi River roiled and roared, swallowing the sandy bluffs as they fell into it, running against the normal current as it rushed to spread back into new holes and crevices, creating waterfalls where they had never been. A comprehensive summary of eyewitness accounts, collected by Timothy Flint, was published in Boston in 1826. Flint wrote:

> The shock of these earthquakes . . . must have equalled in their terrible heavings of the earth, any thing of the kind that has been recorded. . . . Whole tracts were plunged into the bed of the river. The grave-yard at New Madrid, with all its sleeping tenants, was precipitated into the bend of the stream. Most of the houses were thrown down. Large lakes . . . were made in an hour. Other lakes were drained . . . and when the water disappeared a stratus of sand . . . was left in its place. The trees split in the midst, lashed one with another, and are still visible over great tracts of country, inclining in every direction and in every angle to the earth and the horizon. They described the undulation of the earth as resembling waves, increasing in elevation as they advanced, and when they had attained a certain fearful height, the earth would burst, and vast volumes of water, and sand, and pit-coal were discharged, as high as the tops of the trees. I have seen a hundred of these chasms, which remained fearfully deep, although in a very tender alluvial soil, and after a lapse of seven years. Whole districts were covered with white sand, so as to become uninhabitable. The water at first covered the whole country. . . . The birds themselves lost all power and disposition to fly, and retreated to the bosoms of men. . . . A few persons sunk in . . . chasms, and were providentially extricated. One person died of affright. . . . A bursting of the earth just below the village of New Madrid arrested this mighty stream in its course, and caused a reflux of its waves, by which in a little time a great number of boats were swept by the ascending current . . . and left upon the dry earth. . . . They remark that the shocks were clearly distinguishable into two classes; those in which the motion was horizontal, and those in which it was perpendicular. The latter were attended with explosions, and the terrible mixture of noises, that preceded and accompanied the earthquakes, in a louder degree, but were by no means so desolating and destructive as the other.

SUE HUBBELL

Modern earthquake deaths and injuries are caused by the collapse of modern structures. The chimneys of the hundred or so low cabins in New Madrid collapsed as they, and the church and schoolhouse, crumbled, but few people died, even in the boats on the river. Including the person who died of "affright," fewer than a dozen deaths were reported in all accounts.

In the summer, I went to Washington, D.C., and, using the research facilities at the Library of Congress, read through the geological literature on the New Madrid Seismic Zone and tidal-force theory. By also reading the *New Madrid Weekly Record*, the *St. Louis Post-Dispatch*, and my own Ozark hometown newspaper, I was able to follow, with increasing amazement, the growth of earthquake fever.

Every day, the moon exerts a steady, smooth gravitational pull on the earth as it moves in its orbit. On the oceans, where there is no obstruction, it tugs the waters and causes tides. On land, the response is not so dramatic, but there is still a small, measurable bulge, nowhere more than one foot high, pulled along by the moon. When the sun is aligned with the moon, which on December 3, 1990, was at its nearest approach to the earth since 1912, the total gravitational force is only a little stronger than it is the rest of the time. The strain that was put on the New Madrid Seismic Zone on December 3 can be measured and is represented by the number 1.186×10^{-8}. That number is interesting only when it is compared with the strain peak on the New Madrid Seismic Zone on January 17, 1988, another date of high tidal force: the figure then was 1.177×10^{-8}, only slightly smaller. There was no big earthquake on January 17, 1988, either. To put it another way, the strain over and above that during a normal high tide is analogous to the change in air pressure you would feel during a slow ride down in an elevator from the top floor of a ten-story building.

Nevertheless, we have all sat on beaches contemplating the tides, and it seems intuitively obvious that a force that can shift that much water *should* be able to release the chthonic power of an earthquake. Browning's theory of tidal-force triggers is not new. My 1890 *Encyclopaedia Britannica* discusses it, and assigns its "recent" formulation to Alexis Perrey, of Dijon. It is not rare, either: the *Old Farmer's Almanac* states flatly that the five days around high tides are the most likely ones for earthquakes in the Northern Hemisphere. The theory has occurred to scientists, too, and they have studied the correlation of tidal maxima with earthquakes. I called an old friend who is a geologist—Tom Simkin, a volcano expert at the Smithsonian Institution—after I discovered that he had contributed material to an ad-hoc group of eleven geologists who reviewed Browning's "projection" for the National Earthquake Prediction Evaluation Council, an advisory committee to the United States Geological Survey. He sent me a 1973 paper he had co-written pointing up a startling correlation between tidal peaks and earthquakes coming after the collapse of a volcanic caldera in the Galápagos Islands. I began to accumulate other papers that showed positive correlations.

I phoned Brian Mitchell, the chairman of the Department of Earth and Atmospheric Sciences at St. Louis University—also a member of the ad-hoc group—whose specialty is the New Madrid Seismic Zone. I asked him about the studies. "The trouble is that they show a variety of correlations," he said. "Some are with tidal maxima, and some with times in between. A number of them are based on erroneous data or contain mistakes in the calculation of data." He recommended that I talk to Thomas Heaton, still another member of the ad-hoc group, who is the seismologist in charge of the U.S.G.S. office in Pasadena and is a specialist in tidal theory.

"Well, it's hard," Heaton told me when I asked him about the correlations. "There are some late papers that show a correlation with earthquakes and tidal forces and some late papers that don't show a correlation. For instance, in 1975 I wrote a paper that showed that, while for earthquakes worldwide there was no correlation, for shallow earthquakes with a vertical component there was a slight correlation. But then I redid the data and found that under stricter statistical conditions there wasn't *any* correlation. So in 1982 I published another paper, saying, in effect, 'Never mind.' Browning cites my first paper in support of his theory but never mentions my second."

The ad-hoc evaluators had asked Browning about the omission. One of them, Ira Satterfield, from the Missouri Department of Natural Resources, in Rolla, talked about it on the local public-radio station on November 30. He said, "We asked Browning in our telephone conversations, 'Why don't you refer to Dr. Thomas Heaton's latest publication on this instead of his earliest one?' His reply was, 'It doesn't fit my situation.'"

During my talk with Heaton, I asked him if the 1811–12 earthquakes at New Madrid had happened at a time of high tidal forces. He rummaged through some papers, and said, "On December 16, 1811, there was a peak tidal force, but not as high as one in 1809. There was an even higher one in December 1813. January 23 and February 7, 1812, the dates of the big subsequent earthquakes, were not on tidal peaks at all." About all that can be said is that some earthquakes happen on tidal peaks. But tidal peaks happen without earthquakes. They can't be used as predictors.

By autumn, Iben Browning was refusing all requests for interviews, and the man he had referred me to, David Stewart, proved nearly as elusive. "Out" or "Busy" was the answer I got when I telephoned his office, and he did not return my calls. Stewart *was* busy. He was the source that journalists from all over the country and from some other countries as well were going to when, in that newspaperly tradition of evenhandedness, they wanted a quote to counter the uniformly negative comments they were receiving from other geologists on the Browning prediction. In addition to giving quotes to the *Times*, the *Chicago Tribune*, the *Wall Street Journal*, and the *Manchester*

Guardian, Stewart was putting out news releases of his own about earthquake preparedness, talking to groups, and training others to talk to groups. I saw a picture of him in the *New Madrid Weekly Record* briefing Marilyn Quayle, the wife of the vice-president, on the New Madrid Seismic Zone. But in early October his associate director, Michael Coe, who was formerly the head of the earthquake program at the Missouri Emergency Management Agency and had been reassigned to Stewart's office to help with the press of business, did phone me. He was able to answer a few questions, but referred me to Stewart for most of them. Stewart, however, was out of the office; he was visiting Browning. Browning himself had apparently had his fill of interviews; a British production company called TANSTAAFL (There Ain't No Such Thing as a Free Lunch) put out a hundred-minute video that was billed as his final and farewell interview. It had been advertised for ninety-nine dollars in several Midwestern newspapers under the words "A MAJOR EARTHQUAKE HAS BEEN PROJECTED BY DR. IBEN BROWNING TO STRIKE THIS AREA DECEMBER 3, 1990," and the ad featured an endorsement by Stewart: "He is, perhaps, the most intelligent man I've ever met."

In the TANSTAAFL videotape I was able to see that Browning is a corpulent man with glasses, a pale complexion, and short-cropped hair who speaks with assurance as he hands out snippets of highly complex information from such an array of fields that it is difficult for a viewer to evaluate his conclusions. Browning was born in Texas in 1918 and was a precocious child. He graduated from Southwest Texas State Teachers College at nineteen with a degree in mathematics. He was a test pilot during the Second World War and spent his off hours reading the *Encyclopaedia Britannica*. "I read articles at random, integrating them into what I already knew," he told an interviewer. After the war, he accumulated graduate degrees in the biological sciences from the University of Texas, worked as a researcher in a variety of defense-contract companies, and became an inventor, an author, and a business consultant. He is married. His daughter, Evelyn Garriss, who assists him, has an academic background in history. Now, at seventy-three, he suffers from complications of diabetes. He told one interviewer that he would be surprised to still be alive on December 3. (He was.) He received countrywide fame as a speaker at Blanchard's New Orleans Investment Conferences, beginning in 1985. These are annual conferences, organized by James Blanchard, the founder of the National Committee to Legalize Gold, and a conservative who would like to see the United States return to the gold standard. His self-confessed heroes are Ayn Rand and Barry Goldwater, both of whom have spoken at the conferences. Other speakers have been Milton Friedman and, in 1990, along with Garriss, Arnaud de Borchgrave, the editor of the conservative *Washington Times*. The conferences are attended by private investors who don't like to put their money into the hands of brokers. Much of the TANSTAAFL video is taken up with advice from Browning and his daughter about investments in agriculture futures and the like, which are supposed to be made attractive by Browning's insistence that the earth is entering a period of colder weather. The press has routinely reported that Browning is a consultant to PaineWebber, but he has not been one for the past two years. He was one of "fifteen or so" consultants, and the only climatologist, who were used by PaineWebber for about ten years, according to Judith Seime, a research assistant at PaineWebber's Chicago office. She would not venture an estimate of the accuracy of his predictions.

Browning predicts lots of things. In the TANSTAAFL video he says that by 2050 half the world's population will have died from AIDS, and that in a thousand years it will have become an unimportant disease, like measles. In a book published in 1981, *Past and Future History*, Browning has a nine-page list of "inferred" events that will take place between 1980 and 2010. Among them are the following:

Canada will cease to export grain because the climate will get too cold.

Egypt has a 50% chance of having its Aswan High Dam removed by . . . atomic bombs.

Arizona will be wetter than in the last 1000 years. [Right so far.]

Feudalism will sweep the earth.

Eastern Europe will break out of Russian control.

France will be deeply involved in retaking an Empire.

Nineteen eighty-six at 30 degrees south latitude and 1990 at 30 degrees north latitude will be exceptional years for earthquakes and volcanoes. [Wrong.]

Quantitative people will have high enough skills to go to robotic slaves but the humanistic people will turn to human slaves.

The TANSTAAFL video includes an intercut earlier interview purporting to show that Browning did indeed predict the San Francisco earthquake. That tape, according to the interviewer, was made in Albuquerque on September 16, 1989, one month and one day before the earthquake. During that portion of the video, a message in block letters fills the screen: "Dr. Browning warns of earthquake activity on or about 16 October, 1989. San Francisco earthquake occurred 17 October, 1989." But what Browning is actually saying behind those suggestive words is the following: "As of October 16, 1989, there will be a full moon. It's a 413-day precursor high tidal force of the high tide of December 3, 1990. And there'll be earthquakes go off and perhaps a volcano or two. . . ."

There are earthquakes someplace in the world nearly every day. Browning's prediction, which left size and location unspecified, would have seemed impressive only to someone who didn't know of their frequency. Most news reports said that Browning "claimed" to have predicted the San Francisco earthquake, but the word "claim," like the journalistic "alleged," is passed over by many readers. One reporter, William

Robbins, of the *Times*, tried to substantiate the claim. He called a man who said he had heard Browning in a speech forecast the San Francisco earthquake, but the devastating magnitude 7.1 earthquake and the grim TV pictures that followed may have given Browning's words more specificity in the man's memory than they actually had. The ad-hoc group obtained a transcript of the speech and found that he had said only that there would probably be "several earthquakes around the world, Richter 6 plus, and there may be a volcano or two"—little more than he said in the taped interview on the video. Over the past ten years, there has been an annual average of 110 earthquakes of magnitude 6 or greater, working out to about one in any given three-day period. Browning's was a safe forecast, worldwide.

Who, I wondered, had given Browning his pulpit at the Missouri Governor's Conference on Agriculture, where all this started? That person, I found out, was Tom Hopkins, an employee of the Missouri Department of Agriculture and the coordinator of the conference. I talked to him in late October. "Well, this is the way it worked," he told me. "Before I get programs ready for these conferences, I ask around. Bill Galbraith—he's with the North American Equipment Dealers Association—he told me about this guy Browning. He said he'd spoken at one of their meetings about ten years ago and had made a bunch of predictions. Some of the guys kept track and a lot of them came true. Some say he's a quack, though. So I went to my boss, Charlie Kruse"—Missouri's Director of Agriculture—"and his reaction was 'You got to be kidding. We're not going to have some climatologist witch doctor at the conference.' But I convinced him. I said, 'Trust me, Charlie.' I was talking to Dr. Browning just last Friday. I called him to see if we could schedule his daughter to come to speak at this year's conference, on December ninth to the eleventh. He said, 'Well, if what I think is going to happen does happen, you aren't going to have a conference, but we'll book her.' And I said to him,

'There better be an earthquake on December third or when she comes in the room everyone is going to say "Quack, quack."' And he said, 'Tom, what we're saying is going to happen will happen, maybe on December third, maybe January first, but it's going to happen.'"

Academic and government geologists and geophysicists were put in a difficult position by the Browning prediction. They knew that the New Madrid Seismic Zone was active and potentially dangerous; they had been trying for years to alert the public and improve building standards. But they didn't believe that anyone could forecast the day of an earthquake. On the other hand, they couldn't guarantee that there *wouldn't* be an earthquake on December 3. Geology is not a laboratory science; events cannot be repeated, or conditions altered. Geology's time span allows for no such neat experiments. Geologists speak of probabilities for earthquakes over large periods of time. Arch Johnston, the director of the Center for Earthquake Research and Information at Memphis State University, is a New Madrid Seismic Zone specialist and was also a member of the ad-hoc group. He has spent as much time as anyone has thinking about the potential for another devastating New Madrid earthquake, and has worked out the probabilities, on the basis of seismic data and the slim geohistorical record. The geologic evidence shows that there have been some very large earthquakes in the New Madrid Seismic Zone in times past. But, although there are hundreds of small earthquakes there every year, since the series of earthquakes during the winter of 1811–12 there has been only one large one—a IX, in 1895. In working out the probability of the occurrence of such an earthquake, Johnston and his colleague Susan Nava, in a 1985 paper, remind the reader, "The probability estimates . . . rely on the assumption that the New Madrid Seismic Zone generates major earthquakes in a repeated fashion."

I asked Brian Mitchell, at St. Louis University, about this assumption. "It is basic to the analysis of earthquakes," he said.

"It is as good an assumption as can be made. Geologic evidence from trenching through a part of the fault shows earthquakes that have caused liquefaction in the substrate, which indicates that there have been at least two previous severe earthquakes. These cannot be precisely dated but one may have occurred from six hundred to one thousand years ago and the other from six hundred to a thousand years before that."
In the spring of 1990, in plenty of time for anyone who wanted a comparison with the Browning prediction, the U.S.G.S. published a circular, "Tecumseh's Prophecy: Preparing for the Next New Madrid Earthquake," which, using Johnston's work and that of others, estimated with appropriate scholarly caution that the probability that an earthquake of magnitude 6 to 6.5 would occur in the New Madrid Seismic Zone in the next fifteen years was somewhere between 16 and 63 percent.

This isn't fun. This doesn't send a little thrill of fear down the spine. This doesn't get people excited over earthquake preparedness. This doesn't get you on the evening news. What does do those things is the firmed-up Browning prediction that came out of a memo written by David Stewart last summer after an earlier visit to Browning. The memo was addressed to a number of geologists, but it circulated widely in Missouri state offices and in the press. In it Stewart accepts Browning's successful prediction record and praises Browning's abilities. It is here that Stewart makes the comment about Browning's intelligence which was used in the advertisement for the TANSTAAFL video. Stewart admits that he does not understand how Browning picks the particular spot where an earthquake will occur on a day of high tidal force. (According to Satterfield, that turned out to be simple. When asked about it by the ad-hoc group, Browning replied, "Well, like you-all, I read the paper," and went on to explain that he watched to see where geologists were spending money doing research.) In the memo Stewart gave the prediction greater specificity. He said that Browning was concerned about earthquakes on December 3,

1990, plus or minus two days, on the Hayward Fault (in California), in the New Madrid Seismic Zone, and in Tokyo. The memo went on to say, "He would assign a 50% chance to each of these, independent of the other, and in the aggregate an 87% probability that at least one of these three will go December 3. He is virtually 100% certain that some major quake will occur in that band of latitudes on or about that date." Those percentages, unlike the less exciting ones from the U.S.G.S., turned up frequently in news reports thereafter.

Mark Twain never wrote about the New Madrid earthquake, but he did write about the Mississippi River and the kinds of conclusions that the study of it can lead to. In *Life on the Mississippi* Twain described how the river, in its convoluted, curling path, sometimes cut a new channel through an oxbow. Such a cutoff was important to river pilots, for it could save them miles. He documents several cutoffs that he knew of from his own piloting days and a number that had been recorded over the years. And then he writes:

> In the space of one hundred and seventy-six years the Lower Mississippi has shortened itself two hundred and forty-two miles. That is an average of a trifle over one mile and a third per year. Therefore, any calm person, who is not blind or idiotic, can see that in the Old Oölitic Silurian Period, just a million years ago next November, the Lower Mississippi River was upward of one million three hundred thousand miles long, and stuck out over the Gulf of Mexico like a fishing-rod. And by the same token any person can see that seven hundred and forty-two years from now the Lower Mississippi will be only a mile and three-quarters long, and Cairo and New Orleans will have joined their streets together, and be plodding comfortably along under a single mayor and a mutual board of aldermen. There is something fascinating about science. One gets such wholesome returns of conjecture out of such a trifling investment of fact.

SUE HUBBELL

116

My copy of the Stewart memo is from a state office, and someone has underlined a section of it that mentions bridges collapsing on the Mississippi and the military coming in with pontoon bridges. Someone has scrawled "Very interesting" across the top of the memo. Bureaucracies, public and private, fired up their WATS lines and Xerox machines and began holding meetings, conducting earthquake-preparedness seminars and drills, and issuing information to help the public prepare quickly for a destructive earthquake. Schools announced plans to close on December 3. In the autumn, minor tremors along the seismic zone, the sort that would not have rated even a mention in years past, provided the news pegs to hang stories on in newspapers and news magazines across the country. The New Madrid Seismic Zone made *Time*, *Newsweek*, and national television. The United States Figure Skating Association, which was scheduled to begin its regional competition on December 3 in St. Louis, postponed the meet. My hometown newspaper reported that a pair of state troopers, authoritative, shiny, and trim in their uniforms, had given a talk about this man who had correctly predicted the San Francisco earthquake and who now said that there was a 50-to-87-to-100 percent chance of a big earthquake, maybe in New Madrid. They urged survival preparedness for the entire town.

In mid-October, the ad-hoc group's evaluation of Browning's prediction was released at a news conference and widely reported. Using the data Browning had provided for them, the committee members had tested his methods and found them no better than random guessing for predicting earthquakes, or, as one panel member, Duncan Agnew, said, "You could select the dates by throwing darts at a calendar, and you would do as well as Dr. Browning did." They also examined his track record. "This was important," Arch Johnston told me in discussing the report. "Even if his science wasn't valid, if he had accurately predicted seismic events—well, then, perhaps he knew something we didn't." They examined his claim to have predicted not only the San Francisco earthquake but other earthquakes and also volcano eruptions, and found in every case no evidence to support the claim.

In the TANSTAAFL video the interviewer asked Browning why it was that his work was not accepted by government or academic scientists. "I don't have any interaction with the government or academia," Browning replied. "I deal with people who solve problems, not people who make problems." And when the ad-hoc group asked him for his policy recommendations his written reply was "As to policy—I strongly recommend against one. The government has an unblemished record of screwing up everything it touches."

Stories about the ad-hoc group's news conference were followed, on October 21, by a story discrediting Browning's chief supporter, David Stewart. According to the *Post-Dispatch*, Stewart had been involved in an analogous earthquake scare in North Carolina in the 1970s, this one involving a psychic. After the furor over that failed prediction died down, Stewart had been denied tenure at the University of North Carolina, where he taught geology. A few days after the *Post-Dispatch* story appeared, Stewart issued a statement saying he would no longer talk about Browning. He left a message for me with his secretary which said, "There is nothing true about the *St. Louis Post-Dispatch* story of 21 October." But Stewart had cut a swath in the press in North Carolina, and a trail of news stories and interviews seemed to corroborate the *Post-Dispatch* story.

According to these, Stewart, who is a fifty-three-year-old Missourian, grew up hoping to be a minister, but while he was attending Missouri's Central Methodist College he found that his belief in reincarnation clashed with Methodism, and he dropped out of school. He hitchhiked around the country and ended up in an ashram near San Diego. One day, while he was practicing yoga, a message came to him to leave, to marry the woman with whom he believed himself to have been linked in another life, and to have five children. He did these things in the process of building a career and acquiring a Ph.D. in geophysics from the University of Missouri at Rolla. In 1971, he went to teach at the University of North Carolina at Chapel Hill, and became a popular professor. In 1975, he picked up a copy of the *National Enquirer* that told about supposedly successful earthquake predictions made by a psychic named Clarissa Bernhardt. (She also predicted that Nelson Rockefeller would become president in 1976, that Queen Elizabeth would abdicate, and that Johnny Carson would be a blackmail victim.) Impressed, Stewart invited Bernhardt to North Carolina to give a reading of the state's seismological condition. Using an airplane, Stewart and Bernhardt toured the state. Afterward, Bernhardt predicted that the Wilmington area would have a magnitude 8 earthquake within a year of her visit and most likely within a few days of January 17, 1976, a date that Stewart later said he had pressed her for. This prediction badly frightened the area's residents, because the Wilmington area was the site of a nuclear power plant. The earthquake did not happen. In 1977, after Stewart was denied tenure, he told an interviewer, "I have now discovered that science does not even touch the truth. . . . I may or may not remain active in the geologic sciences. Probably not. It was an interim thing in my life. I'll probably spend full time in the childbirth movement." After a tenure-decision appeal was turned down, Stewart ran a publishing company and wrote books on natural childbirth and midwifery. In 1988, he became the director of the Central United States Earthquake Consortium, a group of earthquake-preparedness officials, and later that year he joined the faculty of Southeast Missouri State University. The next year, he became the director of the Center of Earthquake Studies. Three-quarters of his salary was paid by the university and the remainder by the Missouri Emergency Management Agency.

 The evaluation of the Browning prediction and the news story about it had little impact. By that time, the earthquake seemed to have become an imperative. A friend of mine, a teacher in southern Missouri, told me, "I listen to my kids and their parents. They *want* an earthquake. They remember all that TV coverage of the one in San Francisco and it's almost as if they were saying, 'Hey, we could have one of those. Just like the big city. We could be on TV, and all.'" And by autumn many people and organizations had a stake in an earthquake, or at least the fear it generated. Sales of earthquake insurance jumped 50 percent, totaling twenty-two million dollars in Missouri alone. A poll showed that half the population of the St. Louis area believed there would be an earthquake. Discount stores like Wal-Mart laid out displays of earthquake goods, taking as their guide special pull-out sections of newspapers across the region. Based on information from the Center for Earthquake Studies, these told what to do in an earthquake and what supplies to have available. The *Post-Dispatch* published a twenty-page earthquake-preparedness guide a week after it ran stories about the ad-hoc group's denigration of Browning's prediction and about Stewart's credentials. The section was stuffed with ads, including one for guns. The *Post-Dispatch* also ran a food-page special headed "Stocking an Earthquake Cupboard." By late October, you could pick up your free copy of "Tremor Tips" in many public places, including schools. It came from Stewart's center and was Browning-friendly. It contained no ads but listed supplies necessary for riding out an earthquake and told how to prepare the house for one. The Red Cross issued information to smaller newspapers and distributed whistles to children in New Madrid to wear around their necks. Travel agencies in southeastern Missouri were promoting "earthquake get-away weekends," and so was the Convention and Visitors Bureau in Springfield, Missouri, some two hundred miles west of New Madrid. When NBC broadcast a two-part TV movie called *The Big One*, which wildly and deliciously distorted the destructiveness of earthquakes, Midwesterners gathered around their sets to watch, and their anxiety levels ratcheted upward. Critics accused NBC of bad timing, but its timing was nowhere near as bad as that of Missouri's Emergency Management Agency, which scheduled an earthquake drill, "Show Me Response '90," simulating the effects of a magnitude 7.6 earthquake, for December 1 and 2. The drill would make use of emergency personnel statewide, including the National Guard, which would be flying Cobra attack helicopters. State officials insisted that the drill date was just a coincidence, but no one I knew believed that. Arkansas scheduled a similar drill on the same dates, and over in Tennessee plans were made to warehouse the dead and dying in a high-security prison in the western part of the state. The St. Louis area planned drills of its own for December 2, directed from an underground command post, and enlisting the help of, among others, three sixteen-year-olds, who were assigned to struggle with firefighters after they were "rescued," and to scream, in order to contribute to "a general sense of panic," according to the newspaper report.

I left Washington after Thanksgiving to go back to Missouri. Arch Johnston, on hearing my plans, warned me that I would be in greater danger of harm driving between the two points than I would be in New Madrid on E Day. When I got to my farm, I felt as though I had stumbled into the countdown for Armageddon. Everywhere, I heard stories of people buying gas cans, batteries, even emergency power generators that costs thousands of dollars, stocking their pickups with emergency food and driving them out to the middle of their fields. I heard of white roses blooming blue out of season. The Mississippi River was said to be bubbling sulfur. A kindly, grandmotherly woman told me of the Bag 'Em and Tag 'Em Project: an "outfit" (unspecified) was said to have offered plastic bags and Magic Markers to the school in New Madrid so that overbusy undertakers could enlist the help of children to go out, pick up the dead, and label them as best they could. When I reacted with disgust and disbelief, she said, "Well, it *is* just awful, and they turned down the plastic bags. Took the Magic Markers, though." This was country cousin to a myth that floated around St. Louis earlier in the autumn—one about a major utility that was said to have ordered six thousand body bags with its name imprinted on them. And there were other rumors: the Disappearing Hitchhiker, that staple of modern myths, showed up in the Missouri bootheel, dressed in white, silent until he got out of a car; then he turned and said to the driver, "I'm here to warn you of the earthquake." Some reported that he was an angel. Blackbirds were said to be flying backward over the Mississippi River.

The Emergency Management Agency, which had originally been pleased to have the public's attention, admitted that it was overwhelmed. Its operations officer said, "I wish the panic had not gone along with the preparedness," and confessed that the office was now so overworked that it couldn't handle an emergency. The Earthquake Awareness Task Force announced that it had given out the wrong toll-free hot-line number, causing a travel agent (whose birthday was December 3) in California no end of trouble.

The wonder was that anyone was immune to earthquake fever. But I heard wisecracks about the real disaster that was due to come on December 4, when everyone poured out his bottled water. And a ten-year-old friend of mine gave me a rundown on her school's preparedness plans. The local school, unlike others, was steadfastly staying open on December 3, but the children were being drilled, much to my informant's merriment. She giggled so hard she could barely speak when she told me about the kit given to each room with earthquake supplies in it, including a printed "HELP" sign that uninjured children were to throw out in the hallway to alert paramedics to the presence of wounded ones. The local director of emergency preparedness told me with disappointment that despite the fact that she'd put out collection cans, wrapped in yellow paper with a stunning black zigzag and labeled "Earthquake

Disaster Fund," in seventeen places for five weeks, she'd collected only twelve dollars, and one can had been stolen. One Show-Me doubter had contributed two screws and a dead fly.

In an attempt to calm the populace, the state announced that Governor John Ashcroft, accompanied by Agriculture Director Charles Kruse (the man who had reluctantly given the okay for Browning to speak the year before), would tour the more interesting agricultural stops in the New Madrid area on December 3.

Browning had once described what it would be like on December 3. In the speech in which he supposedly forecast the San Francisco earthquake he also said that on December 3, 1990, "that high tide will be terrible . . . it will trigger volcanoes with the morning and evening suns, skies will glow pink."

On the morning of December 3, under cloudy skies, I was awakened by my cat walking on my face. That was no unusual animal behavior presaging an earthquake. He does it every morning. As I fed him and groped for coffee, I remembered that I had to go to an earthquake. I arrived in New Madrid too late to witness the Illinois radio-station crew broadcasting live from the back table at Tom's Grill, but I heard that they had brought with them a 350-pound man and had him jump up and down to try to start the earthquake. News people filled the town. I counted fifty vehicles with satellite dishes and news logos, and then, like Dr. Robertson, who had counted the tremors of 1811–12, I got tired and quit. I looked local, so I was interviewed a lot. "Trouble is," one reporter told me, "everyone who is scared has left town and we can't get any good quotes from the ones who are here." The school had closed. Noranda Aluminum had promised a smoked turkey to any employee who showed up for work. I met the *Los Angeles Times* man. I heard about a reporter from a newspaper in Prague, Czechoslovakia. I looked in at Hap's Bar and Grill, which had begun a daylong Shake, Rattle, and Roll Party at 6:00 A.M. with beer and a bottomless pot of gumbo, and I could see no one who didn't look like a reporter. A young woman asked me to sign a notebook and tell where I was from. By noontime, she had two hundred signatures. By early afternoon, the museum was sold out of T-shirts. Inside the museum, I found a psychologist from California perched on a chair. He looked as though he were bustling even while sitting still. He had come to deal with "pre-trauma stress in children," he told me, and he handed me his press kit. I asked him what he made of the Bag 'Em and Tag 'Em Project. He waved his hand dismissively. "Oh, that was *last* week's rumor," he said. "This week, it's extra embalming fluid for the mortuaries. These children are frightened. They'll tell you they aren't, but I know kids." He called a little girl over. "You know what to do in an earthquake?" he asked her as he threw a yellow rubber duck up in the air. She stared at him. "What's that?" he prompted. "A duck?" she said. "That's right. Just duck," he said, laughing heartily, if singly. He pulled out a big, fuzzy green hand puppet. "This," he said in an aside to me, "is how we transform fear into anger." To the girl he said, "This is Iben Browning. Don't you want to punch him in the nose?" The little girl looked at him with incredulity and pity, that merciless gaze of childhood, and walked away.

Outside, although it was cold and windy, the town had taken on the air of a street party. Some nearby residents had driven into town to look at the media people and their equipment and preserve the scene on their own camcorders. Others stood in knots on street corners and were interviewed repeatedly. Some fought back. Lynn Bock, a town lawyer, strode into a restaurant and, peering through the viewfinder of his camcorder, pointed it at a table full of journalists. "Okay, everyone," he commanded. "Look right into the camera and, one by one, tell me who you are, where you're from, and who you work for." I watched journalists scuttle after Governor Ashcroft and Senator Christopher Bond, who had also shown up. And after a while I drove back home.

The New Madrid Seismic Zone was uncharacteristically quiet all that day, although there had been a magnitude 1 and a magnitude 2 the previous day. They were just routine, too small to be noticed except by seismographs. There were a few largish earthquakes recorded that day—5.0 to 5.5—but they were far away from any place Browning had mentioned in his predictions: New Caledonia, Tonga, and northern Colombia. And nowhere in the world during the five-day period he had allowed himself was there an earthquake of magnitude 6 or greater.

Iben Browning took his telephone off the hook.

There was one earthquake casualty. An eighty-two-year-old woman in Sparta, in south-central Missouri, following the directions in "Tremor Tips," had been removing breakables from a shelf and had fallen from a step stool and broken her hip.

A week later, David Stewart stepped down as director of the Center for Earthquake Studies and asked to be reassigned to full-time teaching.

A neighbor told me, "The next time some government feller tells me to fill up my pickup with peanut butter and drive it out into the middle of my field, I'm not a-gonna do it."

—*The New Yorker*, February 11, 1991.

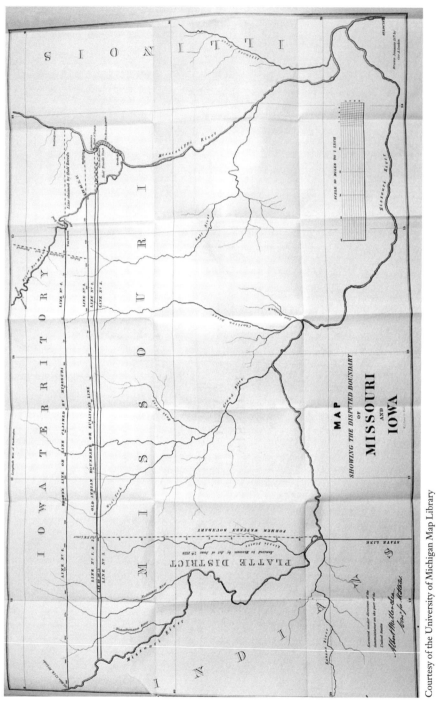

The Honey War

Winter is long and harsh on the prairie. The winds blow down out of the north, running for mile after unobstructed mile, driving snow into drifts around houses, making every human task a struggle. Sometimes men and women go a little mad there in the winter.

The winter of 1839 came early and by the first part of December, the snow was already deep along the Missouri-Iowa border, but on the night of December 15, most of the citizens of Lewis County, in northern Missouri, had braved the snow and cold to meet at Pemberton's Hotel in Monticello, the county seat. They wanted to put on record just how it was they felt about their neighbors—those "foreign bandits"—the residents of the Territory of Iowa, who had "perpetuated a foul indignity upon the escutcheon of our State."

The sheriff from another Missouri county had tried recently to perform his "official duties" in what, they held, was Missouri and had been "forcibly seized and carried away by a band of lawless depredators," the Iowans, an act that threatened "the supremacy of our laws, the inviolability of our soil."

A hot and windy series of resolutions was passed that evening, including one stating

> That when the majesty and authority of a sovereign State has been insulted and contemned by the Governor and authorities of a petty Territory it should seek remedy worthy of itself. . . .

Besides, the Iowans had just called them "Pukes." People out on the frontier sometimes called Missourians "Pukes" and Missourians hated it.

"Death to the Pukes," soldiers in the Iowa army had cried as they marched toward Missouri.

For there was an Iowa army. There was also a Missouri army. Missouri and Iowa were having a war that December. It was as fine a bit of winter madness as the prairie has ever seen.

It may have been the longest war in U.S. history, running, as some authorities will have it, from 1836 to 1851. On the other hand, it may have been the country's shortest, for, in a sense, it stopped before it started.

The casualties were two haunches of deer meat, but they were buried with full military honors.

The war cost Missouri twenty thousand dollars, raised by floating the state's first bond issue. Iowa, still a territory, requested thirty thousand dollars from the U.S. government to pay for its share, but Congress balked and the money was never appropriated.

Depending upon which state one favors, it was called the Iowa War, or the Missouri War. It was also called the Honey War because a contributing cause was the rustling of three bee trees. That act certainly added to the general crotchetiness on both sides, but even winter-weary Iowans and Missourians would not have had themselves a fifty-thousand dollar war if there hadn't been more to it than that.

Basically, it was a border war over the boundary line between Missouri and Iowa and the story started some thirty-five years earlier, once again as winter was coming to the prairie.

In St. Louis, in November of 1804, the United States signed a treaty with the Sac and Fox Indians who ceded their land west of the Mississippi River and north of the Missouri River to a line that runs, roughly, along the present Iowa-Missouri border.

One Indian, however, refused to agree to the treaty. Black Hawk, a formidable Sac warrior with a cool and sober head, pointed out that the chiefs had not been sent to St. Louis to give away land, but to negotiate the release of an Indian held prisoner there. Forgetting their mission, Black Hawk charged, those chiefs "had been drunk the greater part of the time," and returned, much later, "dressed in fine coates . . . and medals."

So Black Hawk and his warriors continued to make life hot for the white men north of the Missouri River to such a degree that the settlers had little time to worry about where the actual boundary of the territory was. In 1815, William Clark, he of Lewis and Clark fame, and by then governor of the Territory of Missouri, issued a proclamation dismissing the Sac, Fox, and Osage claims to the land, declaring, "The pretensions of other nations of Indians to lands lying within these limits being of very recent date, are utterly unsupported by those usages and that possession and prescription accustomed to found their territorial claims." Clark defined the northern "limit" generously, considerably north of the previous Indian line, running from 140 miles north of the mouth of the Kansas River (at the site of Kansas City) east to the Mississippi in a line that ran 10 miles north of present-day Ottumwa.

After the war of 1812 (which Black Hawk joined on the British side) white settlers began to move into the area north of the Missouri River, so in 1816, when Black Hawk finally gave up his opposition and affirmed the treaty of 1804, John C. Sullivan was appointed to survey and mark the northern border of Missouri. Not being quite so greedy as Governor Clark, Sullivan started at a point one hundred miles north of the mouth of the Kansas River and ran the line along what he hoped was a parallel of latitude and marked it by blazing trees, building mounds, and driving stakes. However, "from want of proper care in making corrections for the variation of the magnetic needle," Sullivan's line veered north at the Des Moines River. There were now two more boundaries—the parallel intended and the line surveyed. Both lines were south of Clark's line.

Where was the border? To make matters precisely clear, when Missouri entered the union in 1820, the northern boundary line was described as the parallel of latitude "which passes through the rapids of the river Des Moines."

Clear? Well, not quite clear.

Boatmen on the Mississippi would have guessed the phrase to refer to the Des Moines rapids, marked on old maps, an eleven-mile stretch of rough water a few miles upstream from the junction of the Des Moines River and the Mississippi. A line run through any point in these rapids would have dropped the northern boundary of Missouri to the south of Sullivan's line. But where was the point in that eleven-mile stretch of water from which the line should be run? And was that what was meant by the "rapids of the river Des Moines?"

The questions began to be important as more settlers moved into northern Missouri and southern Iowa. They were a rough and motley crew, remembered by one of their contemporaries, an Iowa veteran of the Honey War, Alfred Hebard, as nomads who lived mostly by hunting, "bushwhackers . . . border ruffians . . . naughty Missourians . . . rude ramblers . . . in a locality where the moral element at the time had few attractions for well disposed people. . . ." For the life of him, Hebard could not see why Iowans would want to settle there. But the area was rich in game and wild bee trees. The latter furnished not only the sole sweetening in the frontier diet, but beeswax, an important item in frontier trade. Bee hunting was business and the land along the border was one of the best places for wild bees. The valley along the Chariton River running some fifty miles north of the present border was even called the Bee Trace.

The "naughty Missourians" and the Iowans were beginning to jostle one another.

The Missouri state legislature ordered the border surveyed by Joseph C. Brown. Instead of trying to find the old blazes and mounds from Sullivan's survey and running the line eastward, Brown decided to use the "rapids of the River Des Moines" as his starting point and shoot a line westward. Ignoring the traditional Des Moines rapids on the Mississippi

River, Brown chose to understand the phrase to mean rapids on the Des Moines River. After wading sixty-three miles up that river, he found a disturbance in the water at the Great Bend. These low-water rapids disappeared after a rain; nevertheless Brown took them to be the rapids referred to in the state's charter and ran his line from them westward to the Missouri River, thus neatly giving the state of Missouri twenty-six thousand square miles more of highly desirable land than did the Sullivan line.

It is no surprise that the state of Missouri accepted Brown's survey, nor is it a surprise that settlers in the area who called themselves Iowans were unhappy with it. The following year, after the Territory of Iowa was separated from Wisconsin, a new commission was established to survey the boundary again. There were seats on the commission for three officials—one from Missouri, one from Iowa, and one from the U.S. government. Missouri, satisfied with the Brown survey, refused to appoint a commissioner, and so the federal man, Albert M. Lea, and Dr. James Brown Davis, from Iowa, began the survey alone. The two men worked through October and well into November, but sickness in the survey party and "the unusually early beginning of a rigorous winter" prevented verification of the full line. Lea, winter weary, concluded that good arguments could be made for any of four boundary lines, but recommended to the U.S. Congress that the old Sullivan line be accepted as a compromise that rather evenly divided the disputed land.

Missouri did not wait for Congress to act, but instead, in February 1839, the state legislature declared that Missouri's jurisdiction extended to the Brown line.

Spring and summer passed. Bees made honey and wax and bee hunters searched out the trees they stored it in. One of the hunters, a Missourian, crossed into the strip of disputed land and cut down three bee trees. Iowans said he had no right to them, no Missouri bee hunters allowed, and sued the rustler in an Iowa court. The judgment went against him and he was fined $1.50 in damages and costs.

In August, Sheriff Uriah S. "Sandy" Gregory, from Clark County, in northern Missouri, went to the only settlement in the disputed land, Farmington, to collect taxes from a group of people who were raising a house. They not only refused to pay taxes, but chased Sheriff Gregory out of town.

Back home, he made his report and it was passed on to Lilburn W. Boggs, governor of Missouri.

Governor Boggs, an uncompromising and opinionated man, is remembered in history for his infamous expulsion of the Mormons from the state and for having given orders to exterminate the Mormons if they did not submit. He was a hardheaded man, both figuratively and literally. One day after his retirement from office he was sitting at home, his back to the window, and a Mormon formerly in his employ tried to kill him, shooting directly at close range into the back of his head. The bullets, however, did not penetrate his skull, and Boggs recovered from the wounds and went on to political glory in California.

Tough, stubborn frontier politician that he was, Boggs could hardly have been amused to hear that a Missouri sheriff had been chased by a bunch of Iowa carpenters.

Later in November, Governor Boggs issued a proclamation urging all officers to stand firm and do their duty and directed Sandy Gregory to go back to Farmington and collect taxes.

This time a large and excited crowd and the sheriff of Van Buren County, Iowa, were waiting for him. The Iowa sheriff briskly arrested the Missouri sheriff and hustled him off to jail in Muscatine, up to the north.

Winter winds and snow had begun to blow across the prairie, and Governor Boggs and Robert Lucas, governor of the Iowa Territory, began raising armies.

Governor Lucas had told his legislature early in November that the dispute might "ultimately lead to the effusion of blood." He was an old hand at bitter border disputes and knew what he was talking about. Before he had come to Iowa, Lucas was governor of Ohio and had led six hundred men to Maumee, where he faced one thousand Michigan soldiers under the generalship of their governor. No one was killed in that, the Wolverine War, only because commissioners from Washington arrived at the last moment and restored the peace. But out across the Mississippi, Washington was far away and Lucas knew just what to do.

Lucas ordered Jesse B. Brown to raise the Iowa militia. General Brown stood six foot seven and had a reputation for high spirits. Once he had dispersed a tavern crowd by throwing into the heating stove a gunpowder keg that he alone knew to be empty, swearing lustily all the while that the population had lived long enough. Brown needed every inch of command and presence to be able to transform Iowa frontiersmen into an army.

A force of one thousand two hundred was called up. Of that number there were 136 officers, ranging from four full generals down to a host of company commanders. But only about five hundred showed up at Farmington for war, a state of affairs understandable when one reads the reminiscences of Hebard, the Honey War veteran who had such a low opinion of the border settlers. Hebard, writing some fifty years after the war, recalled:

> I found, on reaching my cabin . . . an unexpected document, nothing less than a commission from the Commander-in-Chief, appointing me captain of a military company to be raised within a certain defined beat. . . . Recovering from a momentary amazement, I rode over to see and consult with my lieutenants. No matter what we thought, we agreed at once to drum up our beat. Couriers were dispatched to sound the tocsin in the remotest corners, proclaim the imminence of war, and call upon all able-bodied men to appear the following Monday at Billy Moore's blacksmith shop. . . . Also to bring with them or report all war machinery within their reach. . . . At the appointed hour, the Captain, with an old dragoon sword strapped to his side, made a brief speech, saying that all understood the situation as

well as he did, but owing to the great difficulty of providing supplies, equipments and transportation at such an inclement season it was necessary to know first what our force would be. He knew that some could not go, others were disinclined and might risk disobeying orders. To test the matter he scratched a line in the light snow on the ground, and requested all who would go to come forward and "toe the mark." For several minutes no one moved. Presently, however, two sons of Erin, who probably found something somewhere to stir their courage, shoved the toes of their boots up to the line. The infection spread, another and another slowly ventured up, till finally a large majority were on the line, brave and hilarious. The Captain, nolens volens, was "in for it" now. The only thing to be done was to make ready and report as soon as possible. We agreed to meet the next Wednesday and see how near we could get to a starting point. Wednesday came and we straggled together again, but not in a very hilarious mood this time. Many had been painting what they were to leave behind. They fancied a lone cabin in the edge of a grove, with its early smoke rising straight to the clouds; the wood-pile at the door, consisting of a few saplings, half covered with snow, a dull axe leaning against it, waiting to be used; an old cow, with roached back, in the angle of a fence that enclosed the hay, waiting for attention. But where was the man whose duties were thus suggested? He was a hero now marching to the Missouri line, one hundred miles away, to reconstruct the disorderly, while the wife and children and the cow took care of themselves in a temperature below zero.

There was no money to arm the men with guns, so they brought their own weapons including blunderbusses, flintlocks, and swords at best. One man reported for duty with a plow coulter slung around his neck on a log chain. Another brandished a sausage stuffer. S. C. Hastings, later a prominent judge, organized one company, took a position of command behind his men, and ordered them to march, threatening to run the Indian spear he carried through any man who attempted to desert. One small company was organized and equipped with a train of six wagons to carry supplies; five of those wagons were loaded with whiskey.

Such was the army that advanced toward Missouri crying "Death to the Pukes!"

On the Missouri side, things stood little better. There, twenty-two hundred men were mustered from the northern counties. The snow was deep, the weather cold, and the men called had few weapons, no blankets, no tents. Finding store owners reluctant to sell those items to them on the credit of the State of Missouri, some soldiers simply broke into stores and took supplies and groceries. A Missouri veteran recalled what it was like camped out on the Fox River, just south of Farmington:

> About sunset we struck up our fires, for tents we had not, at Camp Hard Times, so called for divers causes. Numbers of us laid by our fires without tents or blankets, for we could not obtain them on the credit of the State. A little after daylight on the 13th, our commander organized the detachment into four companies, 50 men in each, and gave each captain a list of his officers and men. About sunset we took up the line of march, and proceded to Grave Camp.... Our whole detachment appeared with few exceptions, in person or by substitute, all well armed with the arms that nature gave us, there being only about 40 or 50 guns among us.
>
> About the time we got our fires burning, we received information that we would be turned home, which to us was not as bad as taking tartar. About dark our commander was informed that we wished to confer on the two govenors some honors... we retired a short distance... taking with us a quarter of venison that we had the good luck to kill on the way, which we severed in two pieces, and hung up, in representation of the two govenors, and fired a few rounds at them, until we considered them dead! dead! They were then taken down, and borne off by two men to each Govenor, enclosed in a hollow square, with the muffled drum, and marched to the place of internment, where they were interred by the honors of war. We fired over the graves, and then returned to the encampment.

On the morning of the fourteenth, they received orders to return home, but gathered once more to pass a number of pert resolutions, including one that stated that those who had been lucky enough to obtain guns, blankets, and tents had decided to keep them for next winter's war, "as our notice has hitherto been so short."

The troops disbanded, turned their coats inside out, and wandered home, skylarking, burning down fences, and card sharping along the way. They made such a nuisance of themselves that when the grand jury next met, it indicted one hundred of the war veterans for gambling.

While the ragtag troops from both sides had been massing on the border, delegations from Iowa and Missouri had met and drawn up a truce. One Thomas L. Anderson is said to have urged the acceptance of the truce to the court of Clark County, portraying the horrors of war and the blessings of peace with such eloquence that those hearing him wept. If so, he should have been present the night of December 15 when those citizens of Lewis County, directly to the south, met at Pemberton's hotel and passed their resolutions quoted at the outset. They were furious over the troops' return and wanted no part of any truce, condemning instead the "menial and begging policy instituted by certain modern school peace mongers."

With such touchiness still in evidence, it is a wonder that the truce held, but hold it did. And although various Missouri sheriffs continued, without success, to try to collect taxes and were sued in Iowa courts for their attempts, the troops were not called up again.

Both sides wanted the U.S. government to resolve the issue, and during the early 1840s, Congress debated the border question in a desultory way, eventually, in 1844, calling for yet another survey, provided that the act would be approved by Missouri. Missouri, naturally, did not approve, for it had more to lose than gain by a new survey, and it was not until 1845, when Iowa became a state, that the dispute was submitted to the Supreme Court.

Missouri argued before the court for the acceptance of the Brown line passing through the "rapids of the River *of* Des Moines" at the Great Bend in the Des Moines River, and Iowa stoutly urged a boundary on a parallel of latitude that passed through not those "riffles" but instead the "rapids of the River Des Moines" *of* the Mississippi River. It took the court four years to decide that the Des Moines River had only riffles, that the Des Moines Rapids were on the Mississippi, and that the Old Indian Line, the Sullivan line, was the proper compromise to mark the boundary. Then, in what has the appearance of judicial exasperation at the silliness displayed by both parties, the Court directed that the boundary should be marked by iron pillars every ten miles with "Missouri" engraved on the south side, "Iowa" on the north, and "boundary" on the east.

Those pillars, once cast, weighed as much as sixteen hundred pounds and were difficult to put in place. Roads and bridges had to be built to haul them in. In addition, it was nearly impossible to decide where to put them. Traces of the old Sullivan line of 1816 were eventually found, the magnetic error deduced, but it was further discovered that Sullivan's crooked line was not even consistently crooked. So to mark the squiggles in the border, the surveyors drove wooden posts every mile between the iron posts.

It was not until 1851 that the commission finished its job and Henry Hendershot, commissioner, submitted the bill for the work. It was over ten thousand dollars, and Missouri and Iowa each had to pick up three thousand dollars of the tab. One Iowa legislator, reviewing expenses, objected to the $7.12 per diem charge for surveyors, especially since he himself only received three dollars a day. Eying Hendershot, he protested mildly, "Well, Henry, I had lieve help you steal as any man, but I really think you are dipping a little too deep into the public crib."

And so ended the Honey War.

The clash of human events has sometimes been stimulus to artistic creation, and if the greatness or smallness of the art produced has any relationship to that of the event, the Honey War can, perhaps, be judged by the poem that it inspired. It was published in the *Palmyra Whig* (of Missouri) in October 1839, after the bee trees were cut, but before the troops were called up. It was popular, in a modest way, in the 1840s and was meant, according to its author, John W. Campbell, to be sung to the tune of "Yankee Doodle."

The Honey War

Ye freeman of the happy land,
    Which flows with milk and honey,
Arouse to arms, your poneys mount,
    Regard not blood or money.
Old Govenor Lucas, tiger like,
    Is prowling round our borders,
But Govenor Boggs is wide awake,
    Just listen to his orders.

Three bee trees stand about the line
    Between our state and Lucas,
Be ready all the trees to fall,
    And bring things to a focus.
We'll show old Lucas how to brag,
    And claim our precious honey,
He also claims, I understand,
    Of us three bits in money.

The dog who barks will seldom bite,
    Then let him rave and splutter;
How impudent must be the wight
    Who can such vain words utter.
But he will learn before he's done,
    Missouri is not Michigan.
Our bee trees stand on our own land,
    Our honey then we'll bring in.

SUE HUBBELL

Conventions, boys, now let us hold,
   Our honey trade demands it,
Likewise the three bits all in gold,
   We all must understand it.
Now in conventions let us meet,
   In peace this thing to settle,
Let not the tiger's war-like words
   Now raise too high our mettle.

Why shed our brother's blood in haste,
   Because big men require it?
Be not in haste our blood to waste,
   No prudent man desires it.
But let a real cause arise
   To call us into battle,
We're ready then, both boys and men,
   To show the true blue metal.

Now if the Govenors want to fight,
   Just let them meet in person,
For Govenor Boggs can Lucas flog,
   And teach the brag a lesson.
And let the victor cut the trees.
   And have three bits in money.
And wear a crown from town to town
   Annointed with pure honey.

And then no widow will be made,
   No orphans unprotected,
Old Lucas will be nicely flogg'd,
   And from our line ejected.
Our honey trade will then be placed
   Upon a solid basis,
And Govenor Boggs, where'er he goes,
   Will meet with smiling faces.

—*On This Hilltop* (Ballantine, 1991).

The Honey War

Space Aliens Take Over the U.S. Senate!!!

I am standing in one of the most unusual places I have ever been. It is an alcove in the big newsroom of the *National Enquirer* in Lantana, Florida. The newsroom is deep inside a one-story, pebbled-concrete building with huge windows that make it look like an overgrown elementary school. The building is hidden by trees and lush tropical plants, surrounded by a parking lot. The crowded alcove where I am standing is jammed with the desks, computers, and files of the nineteen staffers who turn out the *National Enquirer*'s sibling publication, the *Weekly World News* (*WWN*), the black-and-white supermarket tabloid that has taken newspapering to its most absurd, post-postmodern limits. It is the paper that screams from the rack at the checkout: "WORLD WAR 2 BOMBER FOUND ON MOON!" "BIGFOOT CAPTURED!" "FARMER SHOOTS 23-POUND GRASSHOPPER!"

Eddie Clontz, its editor, sandy-haired, smiley, bouncy, has invited me to spend the day in this alcove. His desk is heaped with green folders that contain clips from hundreds of newspapers around the world, letters from readers, transcripts of telephone calls that are the kernels of *WWN* stories. The folders will be passed around to the reporters, who will then embellish them and translate them into tabloidese. On top of the stack of folders is a rubber dog mask.

"Eddie," I say, "I can't help but notice you have a rubber dog mask on your desk."

"Yeah. I wear it from time to time, but this is my real reporter-waker-upper," he says gleefully as he opens his desk drawer and pulls out the biggest squirt gun I've ever seen. He aims it at Susan Jimison, alumna of the University of Pennsylvania, specialist on Elvis sightings and people who have been held captive in alien spaceships. She is staring meditatively out the window. "Susan!" he calls to her.

She looks up and groans, "Oh, no! Not again!" but Jack Alexander, Indiana University graduate, formerly bureau chief of the *St. Petersburg Times* and city editor of the *St. Petersburg Evening Independent*, chivalrously holds up his word-processor keyboard to block the shot. He has recently helped break the "story of the century"—that five U.S. senators are space aliens.

Most of Eddie's staffers are alumni of Harvard or Bryn Mawr or other good schools. They include a veteran of the *New York Times*, a former Capitol Hill reporter for Newhouse News Service, the retired editor of juvenile nonfiction for J. B. Lippincott. Salaries for established reporters are seventy-five thousand dollars and more. A recent hire, with no tabloid experience, has started at fifty-three thousand dollars, and editors make salaries well into the comfortable six figures. In an unguarded moment, Eddie, himself a tenth-grade dropout and former wire editor of the *St. Petersburg Times* and the *Evening Independent*, once confessed, "We have to pay them a lot because we are, in effect, asking them to end their careers. . . . We're the French Foreign Legion of journalism."

It wasn't until after the death in 1988 of Generoso Paul Pope Jr., the father of the supermarket tabloid, that visitors were allowed into the building. I feel privileged, at the heart of mystery. I turn to Sal Ivone, M.A., University of Chicago, former corporate speechwriter, managing editor of *WWN*. He is a thin, intense, dark-haired, mournful-eyed straight man to Eddie, the ebullient comedian. I ask him if I can meet the alien who, during the autumn of 1992, amused us at the supermarket checkout line by favoring first George Bush, switching to Ross Perot, and finally settling on Bill Clinton. "Yeah, we might be able to arrange it," he says thoughtfully.

Space Aliens Take Over the U.S. Senate!!!

I then ask Eddie to tell me about the story of the century. "Well, we'd run the story about the alien endorsing the candidates," Eddie explains, "and we got a call from a reader and he says, 'Those guys are not the only aliens around. I can't give you my name and don't try to trace this call, but John Glenn's a space alien . . . and . . . and . . . there are MORE!' I said, 'What the hell! Call 'em up. Maybe the whole damn Senate's space aliens.'"

Reporters started calling Senate offices. Many aides responded with incredulity; some hung up. But some didn't. Four senators in addition to Glenn—Orrin Hatch, Nancy Kassebaum, Sam Nunn, and Alan Simpson—came under suspicion; two aides had wit enough to admit the shocking truth. Said Glenn's spokesman, "Okay . . . Okay . . . you found us out, but remember this—mankind is not alone." Nunn's staffer responded, "I'm almost positive there are more," corroboration enough for a *WWN* story. It was published complete with pictures of all five senators and the Capitol. Nunn himself wrote to Eddie: "While I have not been asked to appear on *Deep Space Nine*, your story about me being a space alien . . . generate[d] a significant number of inquiries to my press office. I, of course, answered all the queries telepathically."

What is this? Fiendishly clever political satire? Sal denies it, but there is a certain brightness in his eyes when I mention that such a piece reminds me of the satires I have heard on National Public Radio. Sal is, he confesses, a fan of *All Things Considered*.

"When am I going to meet the alien?" I nag him.

"Y'know we don't let tourists in here unless it's real safe. We've got the landing pad for him out there in the parking lot. Hmmm. Looks like it's pretty full of cars just now," Sal answers, peering out of the window through the tropical foliage.

Do readers believe what they read in the supermarket tabloids? Attitude surveys, for obvious reasons, are hard to conduct, but 20 percent of the respondents to a 1984 Roper poll said the supermarket tabloids were "accurate," and 49 percent said they were not. Lest journalists from the establishment press feel smug, I hasten to point out that a slightly different but comparable Roper poll conducted the same year about the credibility of the media in general found that 24 percent of the respondents trusted newspapers and only 7 percent of them trusted magazines.

Eddie likes to say that he has two hundred million readers every week. That is an estimate of how many people pass through the supermarkets and see his cover. Other, calmer sources claim that the combined readership of the six major national supermarket tabloids, with their assorted emphases on celebrity gossip, health, horoscopes, and the bizarre, is fifty million a week, based on sales, plus the number of readers to whom purchased copies are passed on. But both of these estimates are "tabloid truth"—numbers with a mere kernel of reality. Figures from the magazine industry's Audit Bureau of Circulation (ABC) for U.S. sales over the last half of 1991 (the most recent comparative statistics available) total a little under eleven million for all six, arranged in declining order of sales: *National Enquirer, Star, Globe, National Examiner, Weekly World News, Sun*. If each issue sold is passed on to a couple of other people, total readership may be somewhere between twenty-five million to thirty million each week.

The usual assumption is that the typical tabloid reader is a middle-aged woman who watches a lot of daytime television. However, no in-depth surveys have been done, and Sal and Eddie tell me that *WWN* is beginning to develop a cult readership among intellectuals. "We've got some college fan clubs," Eddie says as Sal drapes me in *WWN* T-shirts featuring the front page of their biggest seller of all time: "ELVIS IS ALIVE!"

The Globe publications, *Globe, National Examiner*, and *Sun*, have always trailed the others. "Pope led the way," Phil Bunton, editorial director of Globe Communications, told me.

Space Aliens Take Over the U.S. Senate!!!

"We've just followed his lead." Globe Communications' offices are showier than the *Enquirer*'s. They are in nearby Boca Raton, nestled into a corporate park; behind the obligatory jungle of plants, the stagy entrance is set off by a waterfall and an outlandish bronze health-club Atlas holding a huge see-through globe of the world.

I talked with Phil in his office at the *Globe* the day before I went to Lantana. In former days, staffers of the competing tabloids drifted from one to the other, ratcheting up their salaries with each move and changing jerseys in the cricket matches the two companies played against each other. American journalists are beginning to replace the rogue British writers who have long dominated the tabloid staffs, and now bowling is the competitive sport of choice. Phil, a portly, pinkly affable Scotsman with sandy hair, is a former Fleet Street journalist who began his newspapering life as a farm editor in the north of Scotland. He came to the United States in 1973 to work on the *Star*, started by Rupert Murdoch. The *Globe*, the eldest of the publications in the Canada-based corporation, was spun out of a Montreal entertainment magazine in the 1960s.

I asked Phil how it was that British reporters and editors have become so successful in American journalism, taking over not only tabloids but some of our most upscale magazines as well. "I think Brits have a talent that has died out in this country, a talent to sensationalize, to give the audience what it wants," he said. "We give them scandal and gossip. I think most American journalists take themselves far too seriously and regard themselves as being crusaders, as if they are really improving the world. We know it's a business."

But he has also been happy to have the protection of the First Amendment. In recent years, no other *Globe* story has attracted more attention than the one that divulged the name of the woman in the William Kennedy Smith rape case. Once she was identified in the *Globe*, the sober media used her name, too. "We got hauled into court," Phil said. "Florida has a rape-victim shield law which we argued was unconstitutional. The local court supported us. It was a First Amendment case."

Other recent court cases have made tabloid editors more cautious, however; all of them have attorneys who vet every issue. The multimillion-dollar suits for slander and defamation by such celebrities as Frank Sinatra, Cary Grant, and Carol Burnett are often settled out of court or, in the end, reduced on appeal. In the famous Carol Burnett case against the *National Enquirer*, the star's 1.6 million-dollar judgment was eventually reduced to two hundred thousand dollars on appeal to the U.S. Supreme Court. And ordinary people, with private lives to defend, fare no better. Nellie Mitchell, a 96-year-old Arkansas woman who was falsely identified by the *Sun* as a 101-year-old pregnant mail carrier, won a judgment of 1.5 million dollars in state court. But the federal appeals judge has ordered the sum reduced, and the case, Phil told me, is still in process.

When the tabloids want a story or a picture, they are willing to pay for it. The *Star* paid 150,000 dollars to Gennifer Flowers for her claim that she had had an affair with Bill Clinton, a story that swirled around the candidate when he was in the New Hampshire primary. "Maybe 20 percent of our stories in the *Globe* are purchased," Phil told me, "but scandals don't actually sell many issues."

Iain Calder, another Scotsman, known to employees as the "Ice Pick," is editor of the *National Enquirer* and president of National Enquirer Inc. He would not grant me an interview. "I do hope you don't think I'm being rude," Calder had purred over the telephone, "but I wouldn't like to be put in the tabloid context." He went on to explain: "Well, of course, we are a tabloid, but I like to compare us to *People* magazine. Do you know last week we ran the same picture of Princess Di that *People* did? Not just like it, but the very same."

The *National Enquirer*, in its logo, states that it has "the largest circulation of any paper in America." It is true that readers of tabloids commonly refer to them as "papers," but

Space Aliens Take Over the U.S. Senate!!!

the ABC considers all six supermarket tabloids, including the *National Enquirer*, to be magazines, and the California court, ruling in the Carol Burnett case, agreed. Checking ABC figures, the *National Enquirer* is fourteenth in U.S. sales, ahead of the eighteenth-placed *People*, to which Iain Calder would prefer to be compared.

The circulation of the *National Enquirer* and the other supermarket tabloids has been in a steady decline over the past several years. Staffers attribute this to a variety of causes, including the recession, competition from lottery tickets at the checkout stand, tabloid television, and the increasingly juicy content of the mainstream press. But the tabloids are also victims of the success of their own distribution systems.

In founder Pope's day, the sales force he employed to put the *Enquirer* into supermarkets handled his tabloids exclusively. After his death, his publications and sales organization, Distribution Services Inc. (DSI), was purchased for 412 million dollars by a New Yorker, Peter Callahan. Callahan was then head of a small company, Macfadden Holdings, containing the remnants of the publishing empire of Bernarr Macfadden, the physical culturist who brought sensational journalism to the masses. In 1990 Callahan also bought the *Star*, for 400 million dollars, and the following year the Enquirer/Star Group was reorganized as a public stock company within Callahan's purview. DSI now distributes lots of other magazines to the supermarkets, not just the tabloids. Globe Communications has a similar distribution setup, and so these days all the tabloids must compete with many other colorful and titillating magazines at the checkout counter.

Pope, the man who brought tabloids into the supermarket, was born in New York City in 1927. He graduated from the Massachusetts Institute of Technology at the age of nineteen. In 1952, after a few desultory years working in the family sand-and-concrete business, at his father's newspaper *Il Progresso*, and even for a short stint at the CIA, he borrowed enough money

SUE HUBBELL

to buy the ailing *New York Enquirer*, a weekly tabloid newspaper with a circulation of only seventeen thousand. He filled the pages of the *Enquirer* with sex, gore, violence, and big headlines ("PASSION PILLS FAN RAPE WAVE") and circulation began to increase. In 1957 he renamed it the *National Enquirer* and began to dream of a readership of twenty million.

By the late 1960s, Pope was looking for sales outlets that would give his paper more potential buyers. He reasoned that nearly everyone spends some time in the supermarket. The *Enquirer* would be a lucrative item in grocery stores, where the profit margin is measured in a penny or two. Today, with a cover price of ninety-nine cents, each copy of the *Enquirer* sold makes the supermarket twenty-one cents, according to a DSI spokeswoman. Supermarket executives were dubious, however, because of the *Enquirer*'s sleazy reputation. Pope cleaned up the content, brought in more celebrity gossip, and introduced health features. By 1971, the year he moved the *National Enquirer* to Lantana, not far from the house he had purchased, he had already used some highly placed influence to help bring supermarket executives around.

In 1983 Pope told an interviewer, "We got Melvin Laird, who was the secretary of defense at the time, to personally take supermarket executives on a tour of the White House where they spent half an hour with President Nixon. The guys who owned Winn-Dixie, which just wouldn't let us in nohow, walked out of that meeting and said, 'We still don't like the *Enquirer* but we've got to put it in now.'"

Today, Laird, who remembers the meeting a little differently, says that Nixon appeared quite accidentally, but he admits that Pope got one of his aides to arrange a Pentagon briefing for supermarket executives during a convention they were holding in Washington. Laird, who spoke at the briefing, had then invited them on the White House tour.

Early in the 1970s, the *National Enquirer* followed *Women's Day* into the supermarkets, elbowing aside candy bars and razor

Space Aliens Take Over the U.S. Senate!!!

blades at the checkout stand, where customers waiting in line would have to take notice. In its best years, a decade straddling the 1970s and '80s, the *Enquirer* could sell five million copies a week, a figure short of Pope's dream of twenty million but still enough to make him a wealthy man. When he died five years ago, his personal fortune was estimated to be 150 million dollars.

Pope, who had loved his work, read every word in his tabloids and worked in the office seven days a week, sometimes in bathing suit and carpet slippers, often with opera playing as background music, watching, perhaps, for the three topics reporters said they were forbidden to write about: the CIA, the Mafia, and Sophia Loren. He had been advised that he could have made even more money had he expanded and diversified. To such advice he replied, according to Eddie Clontz, "Yeah, but you can only eat so many hamburgers."

WWN, which Eddie considers to be the "last great stand for the old-time tabloid," was founded in 1979 by Pope almost as an afterthought. The growing circulation of the *Enquirer* during the 1970s had been so tempting to Rupert Murdoch, the Australian press baron, that he tried to buy the tabloid. Failing that, he began a new American tabloid on the British model, the *Star*, which he produced in color. To meet the competition, Pope was forced to switch to color printing too. Then, to use his idle black-and-white presses, he began *WWN*, filling it with leftover bits from the *National Enquirer*.

Phil Bunton was the first *WWN* editor, but the gossip and celebrity content didn't work in black-and-white any longer, and circulation reflected that. Eddie was hired in 1981. He became managing editor in 1982, editor in 1989. He studied the successful tabloids of the past and carefully crafted *WWN* on their example, looking to the sensational publications of William Randolph Hearst, Joseph Pulitzer, Bernarr Macfadden, and others as models.

In a sense, tabloids—that is, news in condensed form—have

been with us, in oral tradition, ever since people began entertaining one another with ballads and popular stories. A seventeenth-century ballad told of a phantom drummer who threw children out of bed. The November 10, 1992, *Sun* features a story with pictures that begins: "A young family fled their home in horror after a child-hating spirit repeatedly terrorized their little boys!" Printing presses, once developed, were often used for single newssheets, broadsides, to satisfy a universal human desire to read about sensational events. An early broadside tells the story of one Mary Dudson, who swallowed a small snake that grew inside her and finally killed her. In 1924 the *New York World* published a story about a girl who died and was found to be "harboring" a living snake that she had swallowed as an egg. *WWN* of November 10, 1992: "X-RAY SHOWS LIVE SNAKE TRAPPED IN MAN'S STOMACH."

   S. Elizabeth Bird, a cultural anthropologist at the University of Minnesota, became fascinated with this repetition of folk stories, and that led her to write a book, *For Enquiring Minds*, from which the above seventeenth-century examples are drawn. She studied tabloids as folklore and notes their preoccupation with eternal themes: the hero who didn't die (Elvis, JFK, and Jimmy Hoffa are our current favorites); children raised by animals (remember Romulus and Remus?); ghosts (haunted toasters are big right now); monsters (Bigfoot has replaced dragons); flying saucers (the psychologist Carl Jung once traced back through medieval paintings the image of round, healing objects coming from the skies in times of social disruption); fairy stories about princes and princesses (today's definition of royalty has been expanded to include Oprah Winfrey and Elizabeth Taylor).

   The penny press in America was as popular as it was in England. In 1845 the weekly national *Police Gazette*, which Simon Bessie in his classic *Jazz Journalism* terms "a trade organ of Fast Life," showed publishers what a market there was for lurid stories printed in tiny columns. By the turn of the century,

Space Aliens Take Over the U.S. Senate!!!

Americans had given the world a new term, "yellow journalism," to identify the reckless content of newspapers such as Hearst's *New York Journal*, which used the first comic strip, *The Yellow Kid*, the star of which was dressed in his namesake color. But, in the end, the British have the credit for pulling all the strands together and inventing the modern tabloid.

Alfred Harmsworth, later Lord Northcliffe, saw the potential in 1894 of a small-format newspaper with brief, colorful stories and lots of pictures, ones that appealed to women as well as men. By 1909 his *London Daily Mail* and *Daily Mirror* were selling a million copies a day. Harmsworth eyed the American market and told Joseph Patterson of the *Chicago Tribune*, "If the rest of you don't see the light soon, I'll start one [a U.S. tabloid] myself." So, in 1919, Patterson issued America's first true tabloid newspaper, the *Illustrated Daily News*, fifteen inches by eleven inches, with a front page dominated by a photograph of the Prince of Wales, soon to visit Newport, Rhode Island; a back page featuring bathing beauties; and the pages in between stuffed with picture stories about "personalities" in the news. Imitators were immediate and many. Space here permits singling out only one, but it is one that influenced Generoso Pope and Eddie Clontz.

Bernarr Macfadden, who died in 1955 at the age of eighty-seven, is remembered today as a body builder, a health-food advocate, and the inventor of Cosmotarianism, the "Happiness Religion." But Macfadden also had a strong populist belief in the common man and a shrewd business sense; combined, these led him to establish such magazines as *True Story*, which endured to become the nub of Peter Callahan's Macfadden Holdings. In Macfadden's day, *True Story* editors were taxi drivers and dime-store clerks who were ordered not to edit anything out of the confessions sent in by "ordinary" people. In 1924 he began the *New York Evening Graphic*, a tabloid, on the same principle, enlisting, whenever possible, participants to write the news and advising his editors, "Don't stick with the bare skeleton of facts."

Macfadden developed the composograph, a photograph that was enhanced, altered, sometimes posed. With a composograph, for instance, the *Graphic* was able to show the king of England scrubbing his back with a brush inside his own bathroom (tabloid pursuit of British royalty began long before Fergie and Di). In 1927 the *Graphic* brought both embellishment and the composograph into the famous divorce case of a teenage wife, "Peaches" Hennan, and her wealthy, publicity-hungry husband, "Daddy" Browning. I'll let Simon Bessie tell the story: "Peaches' 'private diary' was serialized and countless intimate pictures were printed. Among the numerous composographs were several of the *Graphic*'s outstanding achievements in this line. Daddy and Peaches were shown playing 'doggies' in their boudoir under the headline WOOF, WOOF, I'M A GOOF. Daddy was persuaded to adopt an 'African honking gander' as a pet and was pictured leading the bizarre animal about New York. The picture was captioned HONK, HONK, IT'S THE BONK."

In the end, the Depression and the change in the spirit of America probably killed the *Graphic*, which ceased in 1932. But Eddie also points out that "sensational publications usually had a tremendous, meteoric rise and then a very precipitous fall." He regards it as his challenge to "take a paper in these times and sustain it as a sensational publication. Gossip, myth, mysteries. Those are all a part and so is making it a forum for readers." Hearst, says Eddie admiringly, "was a master of that."

Today's popular folklore in the tabloids is available everywhere—drugstores, supermarkets, newsstands, train stations—but only for one week. Then it is gone. The publishers' own files are not public, and few libraries, aside from the Library of Congress, the Research Libraries of the New York Public

Library, and the Popular Culture Library in Bowling Green, Ohio, receive them. Those collections are incomplete. This means future folklorists may not be able to read that "ALIENS CONSIDER EARTH THE GHETTO OF THE UNIVERSE," may never know that a "BIBLE SCHOLAR SAYS THAT HEAVEN IS FULL," or that a "RARE MAN-EATING KANGAROO KILLS 27 IN AUSTRALIA!"

On the day I visit the *National Enquirer* offices, Eddie invites me to read his mail as he opens it. *WWN* has recently run a story about the fact, new to many of us, that there are baby ghosts. Eddie tells me that the mail has been heavy from people who want to adopt them. And sure enough, letter after letter that we rip open is from lonely, childless couples or single women who lay out their qualifications to adopt a ghost baby. Eddie looks at me, seems touched, a trifle saddened. "We've gotten a thousand letters. Biggest response I've had in a year."

I look around the newsroom at reporters industriously tapping their word-processor keys, peering at computer screens. I feel disoriented. I don't know what all of this means. And as for that alien, it never shows up. Terrible disappointment.

—*Smithsonian Magazine*, October 1993.

Blue Morpho Butterflies

I am caught in a tangle of arms and legs of four other Americans, huddled on the bottom of a rubber raft, pinned among the boulders in the rapids of the Pacaure River on the Caribbean slope of Costa Rica. A good eco-tourist, I tell myself, would have stayed at home and read about this, perhaps donated the price of the trip to some pleasant tropical science center.

Miguel T. Cabrera, a biologist and our guide, has warned us that we might get in trouble on these rapids. And trouble we were in. I could tell because he had muttered, "Oh shit," and told us to get DOWN!!! The Pacaure River, which falls at the rate of forty-six feet per mile, had wedged our raft—and us—between just two of the many exposed rocks. Any moment now I was sure that the river's force was going to flip us and we frail creatures were going to be hopelessly squished against them. But after only the merest of eternities, Miguel pries us free, asks us to resume our perches on the raft's edges, and begin paddling. FAST! We hurtle down the river to a protected cove to assess the damage (one paddle lost) and to stand by in case rescue is needed for the other two rafts and the supply boat.

My raft mates leap out onto some rocks in glee. They have come for the experience of white-water rafting. I follow with weak knees. I am here for the butterflies.

For years lepidopterists have described to me the beauty of neotropical blue morpho butterflies, found from Mexico to South America. I've stared at their representations in books: the males of individual species in brilliant blues, azures, and purples, in contrast to the duller brown of the females. They are big butterflies, some three inches and more in wingspread. Their color is structural, the effect of light passing through the prismatic scales on their wings, which makes their perceived colors shimmer and change. I knew that color pictures could never do them justice. The males typically patrol the edges of forest, such as those along riverbanks, and I have longed to see them, free and flying, glinting in the sunshine.

So when I had a chance recently to visit Costa Rica to look at butterflies, I asked a contact I had made there where I would be most likely to see blue morphos. He asked around and reported back to me that the best place, the place where I would probably see at least three species of them, would be along the Pacaure River. Perhaps I could take a river trip down it. A river trip? That reminded me of floating the calm, beautiful Jack's Fork River, which runs alongside my farm in Missouri. So I started calling around and found an expeditions company that, by chance, was going to take a small group of people down the Pacaure at just the time I planned to be in Costa Rica. I told them I'd like to join that group.

The company sent me a brief description of the river, but I was busy with other matters and only skimmed it, noting that it was a "Class IV" river. I didn't know what that meant, and so a few days before leaving, I stopped into the library and took out a book titled *The Rivers of Costa Rica*, by Michael W. Mayfield and Rafael E. Gallo. My heart sank when I read the definition of a Class IV river: "Long, difficult rapids with constricted passages that often require precise maneuvering in very turbulent waters . . . conditions make rescue difficult." There followed a detailed description of the Pacaure and, worse yet, pictures. One showed a kayaker coming through a set of rapids, invisible in the churning white water except for two hands holding up a paddle. I nearly canceled the trip.

But I didn't, and in the first two days on the river I had seen males of two of the species. *Morpho pleides limpida* had been the most common. I must have seen a dozen or more of the shimmering blue males with their wings rimmed with a dark brown band. Their lazy flap of wings made them easy to watch at the river's edge whenever we stopped to hike back into the forest over mossy boulders to clear jungle pools where we could swim. When I walked into the rain forest for the first time I had a shock of recognition: It was filled with houseplants. This is what happens when your dieffenbachia, philodendron, ficus, and begonia have exactly what they want. Big trees are swathed with carpets of epiphytes and epiphytes upon epiphytes until it is hard to figure out which are the leaves of the original tree. They become layered, intertwining masses of green with leaves in every shape and pattern. The rain forest is the *Goldberg Variations* of green. In a light gap, Miguel had tenderly plucked from the air a butterfly with orange oblong wings, banded in black and yellow, to show us its four walking legs and two short front ones covered with chemical sensors that defined the pretty, dappled creature as one of the family Nymphalidiae, the largest butterfly family in the world. It flew off as he released it. There were no mosquitoes, blackflies, or obnoxious insects in the forest or out on the river. But at one stop, named Fer-de-lance, we all carefully minded our footfalls lest we meet one of those dull-colored, well-camouflaged poisonous snakes. I saw bustling stingless bees, Melipona, everywhere, and out on the river one day we paddled through clouds of Pierid butterflies, white and yellow. At a rest stop, Miguel scooped up and handed to us a tiny, cute, bright red frog. We handed it around: a poison dart frog, whose skin contains a deadly chemical compound used in former times by Indians to tip their arrows. Once Miguel called out to me while paddling, "Suzannah!" and pointed upwards. There, floating above us, was one of the biggest of morphos, *M. amathonte*, his wings perfect azure, unbanded but with a dark brown edge toward his body. I was to see several more later. Even so, that meant I still had not seen the third species, which was rare but was supposed to live along the Pacaure, too.

Now, standing on the rocks in the protected cove, anxiously watching the other rafts successively pin and then break free, I realize that I am looking directly at the spot where that terrifying photograph in the book about Costa Rican rivers had been taken. The book had described it: "[Here] the lower Pacaure produces its toughest rapids, lower Huacas. Even at low water this one hundred fifty yard stretch of ledges is a solid Class IV rapid. At higher water the holes become voracious."

My raft mates and Miguel are waving their arms and cheering as the supply boat finally pulls through. They are busy and miss what I see. It appears to be a shimmering ball flying a few feet above my head, a tiny flying saucer, blue, but successively cobalt, lilac, violet, purple in its iridescence. It plays tricks with my eyes, gleams and slow dances above the water's edge. It is a male *M. cypris*, a midsize butterfly, the remaining species of morpho I had hoped to see.

All morphos flutter their wings in slow, floppy, almost indolent flight, but *cypris*'s distinctive wingbeat, along with its pattern of coloration, creates an optical illusion that makes it appear ball-shaped, which possibly may fool predators as much as it fooled me at first glance. Little is known of *M. cypris* biology, but the Pacaure River is one of the few places where it is seen. Philip J. DeVries, writing in *The Butterflies of Costa Rica*, says, "The sight of this sailing blue orb against the rain forest background is truly one of the most stunningly beautiful in the Neotropics."

Yes. It is now a picture in my head that I will hold until the end of my days.

—*New York Times Sophisticated Traveler Magazine*, May 15, 1994.

The Gift of Letting Go

My ninety-one-acre farm is at the end of seven miles of bad road in a remote part of the Ozark Mountains. Worked by three billion years of seas, sediments, and erosion, the mountains are worn down to gentle, rocky hills, riddled with caves and sinkholes, cut by rivers, and marked off by rough cliffs. They are covered with such a thin layer of topsoil that farming here is a doubtful enterprise. Nevertheless, for twenty-five years, I've built a home here and eked out a living by combining a special kind of farming—commercial beekeeping—with writing. In the course of that time, I've come to appreciate rock, the base of all Ozark beauty.

Rocks in this area vary in composition, size, color, shape, and worthiness as building material: cherty flint; crumbly sandstone in pinks and yellows; eroded, often fanciful, limestone; and heavy, somber granite. A friend hired a local mason to do some stonework, and the man selected his rocks carefully from the creek bed. Some he refused to haul up "because," he said, "they are too pretty to use."

When my first husband and I bought this farm, the cabin on it was faced halfway up on the east side with little rocks set by the previous farm wife. She had harvested them from her garden as they worked up each spring when the ground thawed. That facing was one of the reasons we bought the place, so handsome it was. Interspersed with the stones were artifacts: Royal Crown soda bottle caps, seashells, electric insulators, tiny flowerpots.

After a few years, our marriage ended, but I stayed on and gradually transformed the cabin into a comfortable, even elegant house. Inspired by my predecessor, I often used rock as a building material. Rock was plentiful and free for the labor of picking it up, and I always had to economize. I learned to use other cheap, available materials. I hauled truckloads of oak pallet trimmings from the local pallet factory and made shingles to side the barn and house. I recycled old, often beautiful, windows scrounged from here and there. I glazed and repaired them, and with them extended and opened up the old cabin. I put passive-solar rock floors in the three south-facing rooms. And although I've had help from friends, by necessity I've learned enough small carpentry to build much of this place all by myself. In the process, I have built myself into it, and it has become part of me. Building had to be worked in around the important matter of earning a living, and it has taken me all these twenty-five years to complete the place.

One thing I'd always wanted was a curving rock walkway from the kitchen door to the driveway. For the past five years, when I've found suitable rocks, I've dug and laid them. This year, as I was hauling in firewood from the woodlot for my heating stove, I took a pry bar with me on each trip to chunk out the big flat rocks to finish the job. Laying stone is gratifying work: You see which free rock will look good in the company of an already seated one, dig a custom hole for it to match the contours of its backside, wedge it into place, and then fill the leftover crannies with fine sand.

The walkway is done now, and I am inordinately happy to step out on it. My back ached a little from lifting the stones—some were as big as any I have ever harvested. Now that I am in my sixties, they begin to seem heavier than they probably are, but the ache always leaves in a day or two. I'm not as poor as I was, and I could have afforded hiring someone to haul in the rocks and lay them, but whomever I would have hired would not have done it precisely my way, and besides, I've learned that the process is the pleasure.

A week before I laid the final stone, I had a call from a Realtor informing me that the offer I had made on a property in Maine had been accepted. It is five acres of land beautifully situated on a hill overlooking, but not on, the ocean. It has on it an ugly, boxy little house. But with a daughter-in-law who is an architect and a son who is a contractor living nearby, it will not stay ugly for long. At my age, I can't take twenty-five years to build this next home with my own hands, so I'll be hiring out more of the work and buying more of the materials.

It is in a place where I have been visiting the past few years, and I like the people I've met there. I'm full of plans for the new land, anticipating with pleasure being near family, looking forward to expanding friendships already begun, eager to grow into the thundering, blustery beauty of the Maine coast. And as the years increase, I know my life will be easier, better, living on a paved road within bicycling distance to town, in a house with an oil-burner backup to my wood stove. So mostly I am excited and pleased about this. But there is a scariness and sadness that goes along with the purchase.

The scariness comes from my fear that I've paid too much. I've never bought such a Big Thing on my own before, and I've had sleepless nights running through several futures involving financial ruin. A child of the Great Depression, I've a talent for making up these stories, the most colorful being the one in which I end up as a bag lady sitting on a curb drinking Bardenheier's out of a paper sack. This scene fades with the dawn, but the sadness remains.

Nor am I yet ready to give up this farm, though I know I soon will. I can't afford, in time, energy, or dollars, two places in the long run. I have made friends here who have become as much a part of my life as the farm has been, and I'll not like saying goodbye. I'll miss, achingly, the beauty of these Ozark hills and their seasons, and the vibrant, zestful natural history of a place about which I have so often written. But the hardest thing is facing the reality that a certain part of my life is over. Like many things in life, my decision to buy a new place and eventually leave this one is a combination of good stuff and bad stuff.

The Gift of Letting Go

During the past five years, my second husband (whose presence in my life involves yet a third house and a part-time city existence) and I have been dealing with aging and dying parents. This has made us examine our own future. Those final years can be good or bad depending upon a series of decisions made over time. There appears to be a variety of ways to grow old with grace, contentment, and, yes, even pleasure, but they all involve a freshness of mind. That approach seems to be characteristic of those who welcome new beginnings and, as a consequence, have the gift of letting go.

The unhappy old people, in my experience, have been those who grew too inflexible to open themselves to new experiences, new places, and new friends, and who tried to perpetuate, with increasing frustration and failure, a way of life they had when they were forty. They wait too long to move and are then miserable when they must. They put off decision making until decisions are made for them and the control of their lives has passed to other hands. They are people for whom process has stopped and preservation, cold, binding, and rigid, is all.

I was thinking over all of that as I laid the last course of my rock walkway. I realized that it was my last project on this place. I no longer farm the way I did ten or fifteen years ago; now I make my living more from writing than from beekeeping. I am engaged instead in a sort of metafarming: I've created a ninety-one-acre nature preserve. The barn, house, and outbuildings are finished. Loop trails through the woods are groomed. After twenty-five years of timber-stand improvement, the woodlot is healthy and vigorous. The seventeen hundred pine trees I planted are now a forest.

I feel a sense of accomplishment, of pride, in what I've done, but it has come to an end. Process is all. And Maine, I reflected as I placed the final stones, is full of rocks just waiting to be harvested and put to use.

—*Living Fit*, May/June 1996.

Mustard

> Sharp mustard rime
> To purge the snotteries of our slimie time.
>
> —*The Scourge of Villanie*, three books of satyres, by JOHN MARSTON, 1598

There must be seasons when the province of Alberta is bleak, but my memory of it is otherwise, for once I was there when the high plains were in bloom. It was summertime, the sun was shining, and that Western sky which always seems to go on forever was a brilliant blue. I was driving westward and on either side of the highway, as far as I could see, the fields were full of golden blossoms. In a café where I stopped for coffee a farmer told me that the plant with the golden blossoms in those fields was mustard. "Did you know," he asked, "that mustard is the most popular spice in the world?" No, I didn't. "Most of it is grown right here. We sell the biggest part of the crop in France, but the United States gets a lot, too."

I suspected him of boosterism, but those familiar with world trade agree that the high plains of Canada produce most of the mustard grown in the world. As to its being the world's most popular condiment, that seemed less startling when I realized that everyone I know has a jar or two of mustard in the refrigerator and I remembered all those little disposable packets of mustard that lurk in desk drawers, leftovers from deli lunches. (Jerimiah Coleman, the famous English mustarder, once said in an entrepreneurial way, "My fortune came not from the mustard people eat, but the amount they leave on the side of the plate.")

It also appears that mustard is our oldest spice, for there are hints of its use among human artifacts dating back to 10,000

B.C. Probably its seeds were sprinkled onto meat and its sharp flavor, released as they were chewed, masked the taste of food kept without the benefit of refrigeration.

People have strong opinions about something that has been consumed and added to food for so long. Everyone I know has his own particular favorite among the literally thousands of available mustards and stoutly defends it against all other kinds. There are two mustard museums in this country. One is in Mount Horeb, Wisconsin, presided over by Barry Levinson, an attorney who wasn't having much fun lawyering. He displays approximately two thousand varieties of mustards; sends out a mustard newsletter to forty-thousand people; and started a club called Mustard Buddies to put fans in touch with one another, which has led to romance, even marriage, between mustard mates.

The other museum is a working one in the easternmost town in the United States, Eastport, Maine, operated by Nancy Raye, a third-generation mustarder. The mustard-curious are welcome to tour the factory there, which is special because most mustard works do not allow visitors and the industry is veiled in secrecy. For instance, RJR Nabisco, which now owns the domestic rights to Grey Poupon and produces it in California, keeps even its recipe locked in a safe at all times, and it is reported that no one ever speaks of it. But Nancy Raye kindly took me on a tour of her factory. Her plant is unusual because it still produces mustard the way it was made for centuries on the European continent, using the stone grinding wheels her grandfather started with. There is only one other stone-ground mustard made in the world, and that is in Germany.

The paired stone grinding wheels, each one weighing a ton, process a slurry of crushed, winnowed mustard seed, vinegar, turmeric, and other spices appropriate to the particular mustard the factory is making that day. The wheels rumble like distant thunder and grind slowly to avoid any buildup of heat.

The making of foodstuffs is, after all, on some level, the

practice of chemistry, and so in order to be a proper mustarder or a proper eater of mustard, one needs to know a little practical chemistry and a bit of botany. There are many kinds of food plants allied to the mustards, genus *Brassica*, among them rutabaga, pak choi, rape, and kohlrabi. There are also several kinds of mustards that produce most of the seeds for the condiment. Popularly these are called black, brown, and white (or alternately yellow) mustards. Each of them has a different flavor, owing to a slightly different chemistry. Black mustard is hard to harvest with modern machines, and the product made from it is runny, needing a lot of cereal as stabilizer. So true black mustard, that is, *Brassica nigra*, is seldom used today in commercial mustards. The seeds sold as "black" are usually from one of the botanical brown mustards or some hybrid thereof.

The flavor of each kind of seed does not become apparent until the husk is broken, its contents crushed and mixed with moisture to trigger an enzyme within the seed which reacts with an internal glucoside, a plant carbohydrate-bearing compound which releases a volatile oil. The oil, in the case of brown mustard, is the staple of Asian cookery. Heat destroys mustard's bite, so this chemical process needs to take place without heat (and that is the reason mustard should be added toward the end of cooking). The piquant flavor of black or brown mustard seems to vaporize in the mouth, producing a horseradish-like effect. The chemical process in the breakdown of white mustard is the same but has different components, and so white mustard's effects are only on the tongue. It is more pungent, but milder. In different parts of the world there are cultural preferences for these different effects. Specialty mustards blend them with spices, wine, honey, vinegar, maple syrup, and a host of other ingredients to create subtleties of flavor.

English mustard, for instance, is a blend of white and brown seeds, turmeric, coriander, and other flavorings. Long ago, in England, mustard seeds were served on the table along with pots of salt and pepper and the diner was supposed to crush the seeds with a knife handle on the table and sprinkle them on the food. And although "paste" mustard is found in England, the traditional English way to serve it is dry but crushed, to be mixed with slightly acidified water and allowed to stand for ten minutes to develop its flavor.

Cold-processed mustard needs to be aged to take out bitterness and to come to fullness of taste. Nancy Raye and her staff sample each batch as it is produced and usually age it for a couple of months. There are surprises. Once, a few years ago, an entire day's production flunked the taste test, and she told her shop foreman not to bottle it. It sat in a storage tank for a year and when the foreman needed the space he asked her what to do with it. "Probably throw it out," she told him. "It tasted awful, but let's taste it again to make sure." When they did they found that it had developed into one of the best mustards they had ever produced, and they bottled it and sold it as a limited-edition house specialty. They have never produced its like again.

The various species of mustard plants, also prized for their greens and a host of medical properties, are probably native to the Mediterranean basin and the Far East. Pliny, that wonderful collector of natural-history lore at the beginning of the Christian era, indicated that the greens were simply collected wild. But there is good evidence that the plants were cultivated for their seed by that time because Romans were mixing mustard seeds with vinegar to make a sauce. Columella, who wrote of Roman domestic and agricultural matters, noted down this recipe around 60 B.C. It is close to a modern one.

> Clean the mustard seed very carefully. Sift it well and wash it in cold water. After it is clean, soak it in cold water two hours. Stir it, squeeze it, and put it into a new or very clean mortar. Crush it with a pestle. When it is well ground, put the resulting paste in the center of the mortar, press and flatten it with the hand. Make furrows in the surface and put hot coals in them. Pour

water with saltpeter over these. This will take the bitterness out of the seed and prevent it from molding. Pour off the moisture completely. Pour strong vinegar over the mustard, mix it thoroughly with the pestle, and force through a sieve.

The Romans were mustard fans and carried it everywhere with them in their baggage trains as they invaded European lands, and went to England. After the collapse of the empire, during those dim years we call the Dark Ages, the French developed any number of local mustards. But throughout the era, Dijon, in Burgundy, preserved the Roman way of making it and became famous for its manufacture. Its popularity there is proved by a feast given in 1336 by the Duke of Burgundy, after which it was noted that the guests had consumed seventy gallons of mixed mustard.

At around the same time, when the popes were in residence in Avignon, a French pope was chosen. John XXII (1316–34) is regarded in official Church histories as a good enough pope and a modest man with only a few worldly weaknesses. Two of them have been recorded. Although there is no hint that he used his office for personal gain, he did like to see his relatives well-fixed and, Frenchman that he was, he liked a tasty, well-garnished meal. Combining the two, he created for his nephew the post of Moutardier du Pape. The nephew became so puffed up with the grandeur of his post that he became the laughingstock of Europe. The joke crossed the Channel, and to this day the term "the pope's mustard maker" is used to describe in both languages a person who has grown pompous over a trifle.

In 1630, the French issued regulations governing the production of mustard, decreeing that it must be made of "good seed and suitable vinegar," without any binder. Later, by edict, spices were allowed. And in the same year, Dijon issued its own rules regulating mustard makers, stating, among other things, that they must wear "clean and modest clothes."

John XXII was not the only head of state to take an interest in mustard. A few centuries later, Louis XIV, the Sun King, gave mustard its own coat of arms: funnel argent on azure. Over in England, Mary Clements of Durham, in the early eighteenth century, developed a new process of fine-milling mustard with such good flavor and appearance that it caught the royal eye and palate. In fact, so pleased with Mrs. Clements's mustard was George I that he bought her business and thus could add Mustarder to his other grand titles.

Historically, Americans used little mustard as a spice. And in some circles at least, mustard was regarded with deep suspicion. In the nineteenth century, Sylvester Graham, a charismatic Massachusetts Presbyterian clergyman, sometimes called the Mad Enthusiast, became founder of a dietary movement based on the eating of whole-grain flour. His name lives on in the cracker. In 1830, he published a tract on the evils of certain foods and roundly condemned mustard, asserting that it injured health and stimulated carnal appetites.

Apparently European mustards were too strong and too hot for American tastes, for it was not until 1904, when the French Company began producing its Cream Salad Mustard, a mild mustard made from white seed, that our countrymen and -women began eating it as a condiment. Even in recent years what we think of as American or ballpark mustard, bright yellow from turmeric and tart from vinegar, accounted for 75 percent of the mustard eaten in the U.S.

This is not to say that mustard, in other forms, was not being produced here earlier. The first U.S. spice-grinding company was opened in 1822, in Boston, by William Underwood, who had learned his trade in his native England at a shop specializing in preserving agents. He ground imported mustard seed in dry form and undersold it in American fishing ports, driving out of business other importers of ground mustard. In 1846, he began packing his own fish and shellfish in Maine.

Eastport, Maine, in the late 1800s, became the sardine-packing capital of this country, and that is why J. W. Raye started his business there. But mustards produced in those days were used almost exclusively as a preservative for canned fish and sardines.

In recent years specialty mustards have become more popular. Nancy Raye estimates that only 20 percent of her production goes to sardine canneries now and the remaining is of gourmet specialty mustards. Americans, like people in the rest of the world, have grown to love mustard. Some, perhaps, might even agree with an ancient English jingle:

> From three things may the Lord preserve us
> From varlets much too proud to serve us;
> From women smeared with heavy fard, good grief!
> From lack of mustard when we eat corned beef.

Not only has mustard become a part of the American diet, but it has also entered our vocabulary. Witness "He puts a lot of mustard on that ball," referring to a fast pitcher. And "proper mustard" means something genuine. But the most curious Americanism relating to mustard is the phrase "can't cut the mustard," meaning incompetence. It is curious because professional derivationists shrug their editorial shoulders, take a stab or two at lame explanations, and then conclude that the derivation is unknown. My own favorite explanation, however, comes from Nancy Raye, who says with assurance, "Oh, it is a term left over from early days when several generations lived and worked the same farm. When Grandpa got too old to work the fields, the family would send him out with a sickle to cut the wild mustard that grew up as a weed in the rich soil around the barn. When he got too old to do that simple task, too old to cut the mustard, they knew he was at the end of his days."

—*Copia*, October/November 2001.

Ozark Springtime

January 15. Snow on ground still from last week's storm, but overhead maples in bloom, their tiny red flowers an item of interest to the bees who are working them for springtime's first nectar. In hives queens will be starting to lay eggs. I badly need springtime myself. Dress in insulated coveralls, mittens, woolen hat, go looking for it. Through snow, through pine grove, past end of farm into U.S. Park Service land to glade on high bluff where Barn Hollow Creek and Jack's Fork River join below. Scramble down the south face of bluff to creek and there they are: clusters of harbinger-of-spring, small white flowers growing from bulbous root. Climb the seventy-five feet back up the limestone bluff, find my favorite cove in the rocks, peel off coveralls, hat, mittens, shoes, socks, sweatshirt. Snow across hollow on north face of cliff, still winter there, but here on south face I curl up and take the first springtime nap.

February 11. Peeper frogs chorusing, courting out in pond. Pumped up with hormones, randy. Soon other frogs will start singing, too, eventually jug-a-rum of bullfrogs and then every puddle, every pond will be full of fist-sized translucent egg masses, promise of tadpoles to come. SPRING!!!

From my farm journal, when I lived on my Ozark farm, seven miles north of Mountain View, Missouri.

The old man drove up my driveway one sunny day in one of those exceptional springtimes when the dogwoods and redbud trees were in bloom during the same week, turning the woods everywhere into a shimmering display. I wasn't expecting a visitor, noted his car had Oklahoma license plates, walked out the kitchen door to find out what had brought him here. He explained that he'd been born "back yonder." He waved his hand toward my open field. He said his folks had bought, in the late 1800s, a piece of land from the family that had homesteaded the place. They'd built a house there and his father had cut timber for the sawmill powered by the flow of the Jack's Fork River. They'd moved away in 1909, and had never come back. "I was just a lad when we left, but I'm past ninety now and it seems every springtime I get a hankering to come back and see the old houseplace," he said.

I'd heard of his family, knew where the place had been. It was on what had become part of the Ozark National Scenic Riverway—U.S. Park Service land. The Park Service had bought, in the 1960s, land all along the Jack's Fork River (once described by a boater as the Mozart of rivers) to complete the eighty-thousand-acre park, the first of the U.S. Scenic Riverways. The old man's home was deep in that park, now an oak-hickory woods.

I told him I knew where it was, had been there a while back, and the rock-lined spring that must have served as the family's well had been full of masses of frogs' eggs. He remembered the well, said his daddy had lined in those rocks. We drove back in my pickup and I showed him the old well, cellar holes nearly filled with leaf mold, iris that occasionally bloomed where they could find a sunny gap in the forest shade. He looked dismayed. "Guess you are right, ma'am. I remember my mother sure did like her flowers, but this was all open. We had a garden over that-a-way; cut hay in a field on the other side."

After a while we drove back through the grove of native white pines—seventeen hundred of which my first husband, our son, and I had planted as seedlings the year we bought the farm in the early 1970s. The pines had grown fast. The wind now soughed through their tops; the roadway was blanketed with pine needles. The old man was thoughtful, quiet. When we got back to his car he said, "Thank you ma'am. It sure is pretty hereabouts, but I feel kinda like a ghost."

March 16. Rue anenome in bloom in shady woods at old girls' camp down on river. At land's edge river flow has laid a big gravel bar. Sweet clover coming up in it, some young sycamores, too. Next flood of the Jack's Fork will uproot them, carry them away, re-arrange the gravel. Sit on bar, admire cliffs, watch buzzards circle overhead in updrafts.

March 24. Pussy toes and tiny yellow flowers of fragrant sumac bloom around house. Cardinal and chickadee songs have long changed to springtime courting tunes. Down at girls' camp blood root, hepatica, Dutchman's breeches have taken over shady woods where camp buildings once stood, long bulldozed away by U.S. Park Service.

April 10. Woods bright with serviceberry trees in full bloom. Sit on top of Pigeon Hawk Bluff to admire them, watch river rushing along far below. One hundred and fifty feet down? Maybe more. Big boulder down there in deep water at river's bend. Called, logically, Rock-in-the-Hole. Pair of tourists floating by it in canoe probably don't know that. Everything has a name in the Ozarks. Back of barn wild plum trees in full bloom, their fragrance like the taste of cherry Lifesavers. Bluebells in bud near river.

I've lived all over the country—Michigan, California, Texas, New Jersey, Rhode Island, and, now, Maine—but I never understood springtime until I spent twenty-five years farming in the Ozarks. Like the old man, I yearn for it every springtime. Spring starts in the Ozarks in January, lurches on in a complicated way with spurts and setbacks until May. There is a cold spell early in May, known as Blackberry Winter because it comes when blackberries bloom. It is a worrisome week for anyone who farms. I remember one particularly difficult year when a killing frost came on May 5, blasting even the new young oak leaves; nothing went well agriculturally the rest of the year.

The complexity of an Ozark springtime is determined in part by the region's geology, in part by contending climatological forces that meet there: prairie's lingering winter contending with the soft, wet ocean air sucked up from the Gulf of Mexico. The Ozarks are old and worn mountains from the geological past. Local people say, "Our mountains may not be high, but our valleys sure are deep." Those deep valleys, eroded by rivers such as the Jack's Fork or Barn Hollow Creek, cut deep box canyons, their south faces protected, warm, a season ahead of the north faces. Updrafts and downdrafts are born from them, giving soaring birds a free ride and fostering change to weather patterns above.

You have to take springtime on its own terms in the Ozarks: There is no other way. It can't be predicted. It is unsteady, full of promise, promise that is sometimes broken. It is also bawdy, irrepressible, excessive, fecund, willful. Asexual beings bud, swell, make spores. Everything that pairs courts, mates, gives birth. Farmers' fields are full of calves, colts, kids, piglets. The woods are full of fawns.

April 11. On wet rocks down in Barn Hollow there is something that looks like goose down, escaped from a pillow. White, filmy, each filament wafting in the breeze. It is a fungus. Grows across rocks. Its filaments, silky to the touch, tipped in gray. WHAT IS? Many bare places in woods where wild turkeys scratched away dead leaves looking for grubs. Tom turkeys beginning to look dazed, ignore me. Hormones drive them to think increasingly *Hen. Hen. Hen. Hen* and nothing else. Makes them rash. Soon hunters will shoot them, tricking them to come running up with even the clumsiest imitation of the hen's sexy *putput . . . putterput.*

One springtime recently I gave in to my yearning and went back for at least a bit of Ozark springtime. I arrived late at night and stole back out to the place where my farm had been. The moon was full and early dogwood bloom made the woods seem to shine. There were moon shadows everywhere. I was too early for whippoorwills—they come in May—but I could hear the *hoohoohooaw* of a barred owl nearby, nearly drowned out by the songs of frogs at their nighttime pleasures. I had known the trees on those ninety acres, some of them intimately, but all around new ones were growing up. What were those new shapes? What were those new shadows? I understood what the old man from Oklahoma had meant when he said he felt like a ghost.

The reason I could prowl around out there in the moonlight was that my piece of land is now public. It belongs to you and you and you and still a little bit to me, too. Barn Hollow is now a Missouri Department of Conservation Natural Area, prized for its unusual geology (five caves, recently mapped) and its rare plants, some of them prairie plants, relic ice age growth. My farm had been the last private land, wedged between that Natural Area and the Ozark National Scenic Riverway, and so, when age told me it was time for me to give up farming and move, the MDC offered to buy the land from me. For a while, after we had arranged the sale, a Conservation Department biologist and his wife lived in my house, used the outbuildings. But when he took a job in another state, the house and buildings were put out to bid for materials and the remaining debris was bulldozed and buried. The Barn Hollow Conservation Area has added 252 acres to the ONSR's 80,000, and is all wild habitat, public land.

Friends had approved the change in the land's status, but lamented the loss of the house. I did not. I'd had my fun building the place. Process is all and the process had ended. It seemed a privilege to be able to return to the wild a place I'd loved for all the wild lives being lived out on it. How often does a person have a chance in a lifetime to give something back to nature?

April 14. Many yellow flowers: fields full of wild mustard, dandelions. In woods: bellwort with its trumpet-shaped hanging blossoms, yellow corydalis, yellow honeysuckle. Throughout woods redbud trees, vibrant, tarty pink in color.

April 22. Dogwoods in bloom. Also May apples, wild phlox, larkspur, fire pinks in woods. Roadsides carpeted with violets. Wet places: lots of buttercups. Spotted a parula warbler. Warblers' migration in full swing now.

I hadn't been back to Mountain View for a number of years because I wanted to wait until the scars from bulldozing had healed before I looked at what had been my farm. When, in daylight, I saw it, it was a place returning to its own nature. It is now like an illustration from an ecology textbook. The land the Park Service has owned since the 1960s, back where the Jack's Fork and Barn Hollow meet, hasn't been logged since the 1940s. The woods there are easy to walk through, clear and free of intrusive underbrush and brambles. They are home to a stout population of wild turkeys and other animals. The transition zone is the one which was my woodlot, buffered by the pine grove. Multiflora roses, blackberries, and a few other invasive exotic plants are being shaded out there, but it, too, soon will be a mature forest. In the open field, past two groves of persimmon trees, a riotous contest is taking place: The white pines' offspring are invading the open land, but so are native cedars, sassafras, and sumac. Blackberries, multiflora rose, thistles, burdock contend for sunshiny places under the young trees. But, in time, that will be wooded, too.

Down in Barn Hollow beaver had built a house in a wide place where the creek pools. They'll soon change its flow to their liking. A young chipmunk, who had never seen a human, sat down a foot from me and tried to figure me out. When I shifted position, he wisely scuttled. The cheerful lobed green leaves growing perpendicular to the cliff walls promised columbine blossoms in early summer. A favorite stand of equisetum, a primitive plant, was still there, reminding me that some species can last for aeons.

I climbed down to the Blue Hole on the river. It is just above the old girl's camp and is, arguably, the prettiest spot on the Jack's Fork River. It is at a turn in the river, a wide, deep pool, its depth so great that swimmers attempting to measure it come up gasping, the bottom never reached. Local people sometimes call it the Gar Hole because those big, toothy fish live in it. I sat on the gravel bar beside it, looked in vain for green and great blue herons. Probably too early for them. I had asked the U.S. Park Service for their updated bird checklist for the ONSR. The area is part of the great Midwestern flyway—the principal air road for migrating birds. I counted nearly two hundred birds on the list. Excluding a few of the ducks, I'd seen nearly all the birds on the list at one time or another during my years on the farm.

The woods around the Blue Hole and the girls' camp below it were carpeted with wildflowers. I walked back up the eroded trail that at the turn of the twentieth century had served as a road to the girls' camp. It is undrivable now, but makes a good hiking path to the river.

Back at my former houseplace I could hear, high in the air, the *kyr kyr kyyyr* of a red-tailed hawk. A yellow-bellied sapsucker was working the ancient oak tree I used to sit under with my morning coffee. The sapsucker was drilling holes, gorging on the rising sap and small insects. Near where the house once stood some iris, which long ago I had transplanted from the old man's houseplace, were just starting to bloom. Eastern bluebirds were busily gathering dead grasses and small twigs for nesting in the hopeful sunshine.

SUE HUBBELL

May 10.   Morning coffee under big oak tree every day now. Stop to check on the spiderlings. Four days ago baby spiders hatched on web on winter woodpile. Hundreds of them. Too tiny to identify, but their web says their mother was an orb-weaver. Each day they disperse in early sunlight into stacked wood. Gone all day. Each morning they are back, must come back in night. Why? Daylight safety? Nighttime warmth? Must find someone who knows spider behavior and find out.

May 14.   Too leafy now to identify warblers except by song. Blue-eyed grass in bloom. Think I'll name the new blue-eyed kitten Rinca after the grass's Latin genus: *Sisyrinchium*. Larkspur, wild hyacinth, wild indigo, wild geranium, horsemint, even ox-eye daisies now. Indigo buntings back livening things up with their tuneful repeating couplets. Whippoorwills. Can't keep up with it anymore. TOO MUCH. TOO MUCH.

—*New York Times Sophisticated Traveler Magazine,* March 2, 2003.